AF801416

Performing Landscapes

Series Editors

Deirdre Heddon, Theatre Studies, School of Culture and Creative Arts,
University of Glasgow, Glasgow, UK
Sally Mackey, The Royal Central School of Speech and Drama,
London, UK

Performing Landscapes offers a critical study of generic and complex sites for performance, including forests, ruins, rivers, home, fields, islands and mountains. Distinctive to this series is that such landscape figures will be located both on and off the theatrical stage, approached as both material and representational grounds for performance-led analyses. With its unique focus on particular and singular sites, Performing Landscapes will develop in novel ways the debates concerning performance's multiple relations to environment, ecology and global concerns.

Editorial Board
Professor Stephen Bottoms (University of Manchester)
Professor Una Chaudhuri (New York University)
Dr. Wallace Heim (independent scholar)
Professor Carl Lavery (University of Glasgow)
Professor Theresa J May (University of Oregon)
Dr. Paul Rae (University of Melbourne)
Professor Joanne Tomkins (University of Queensland)

More information about this series at
https://link.springer.com/bookseries/14557

Susan C. Haedicke

Performing Farmscapes

palgrave
macmillan

Susan C. Haedicke
Department of Theatre and Performance Studies
University of Warwick
Coventry, UK

Performing Landscapes
ISBN 978-3-030-82433-4 ISBN 978-3-030-82434-1 (eBook)
https://doi.org/10.1007/978-3-030-82434-1

© The Editor(s) (if applicable) and The Author(s) 2021
This work is subject to copyright. All rights are solely and exclusively licensed by the Publisher, whether the whole or part of the material is concerned, specifically the rights of translation, reprinting, reuse of illustrations, recitation, broadcasting, reproduction on microfilms or in any other physical way, and transmission or information storage and retrieval, electronic adaptation, computer software, or by similar or dissimilar methodology now known or hereafter developed.
The use of general descriptive names, registered names, trademarks, service marks, etc. in this publication does not imply, even in the absence of a specific statement, that such names are exempt from the relevant protective laws and regulations and therefore free for general use.
The publisher, the authors and the editors are safe to assume that the advice and information in this book are believed to be true and accurate at the date of publication. Neither the publisher nor the authors or the editors give a warranty, expressed or implied, with respect to the material contained herein or for any errors or omissions that may have been made. The publisher remains neutral with regard to jurisdictional claims in published maps and institutional affiliations.

Cover credit: A *Field of Wheat: Arts and Agricultural Project*. Branston Booths, Lincolnshire, England. September 2016
Performer: Anne-Marie Culhane
Photographer: Nathan Gibson

This Palgrave Macmillan imprint is published by the registered company Springer Nature Switzerland AG
The registered company address is: Gewerbestrasse 11, 6330 Cham, Switzerland

Series Editors' Preface

The *Performing Landscapes* series provides an international platform for the first comprehensive critical study of generic but complex sites of performance landscapes, located on and off the theatrical stage and within and beyond the frame of cultural performance practices.

Acknowledging and engaging with the nature-culture dynamics always already at play in any concept of and approach to 'landscape', authors' original research and innovative methods explore how landscapes—such as mountains, ruins, gardens, ice, forests and islands—are encountered, represented, contested, materialised and made sense of through and in performance. Studies of singular landscape environments, experienced from near and afar, offer up rigorous historical, cultural and critical discussion and analysis through the dynamic and interdisciplinary lens of performance. In the context of the twenty-first century climate changes the series also directs attention to performance's diverse contributions to environmental debates.

Performing Landscapes aims to understand better how specific landscape locations function as sites of and for performance and what performance practice and analyses does to and for our understanding of, and engagement with, landscapes.

Professor Deirdre Heddon
University of Glasgow, Glasgow, UK

Professor Sally Mackey
The Royal Central School of Speech and Drama
University of London, London, UK

Acknowledgements

Many people made this book possible from the artists whose work inspired me and audiences who enjoyed it alongside me to scholars and colleagues who critiqued my ideas and made valuable suggestions. A very special thank you goes to Kristen Oshyn, the artist whose evocative pen and ink drawings bring a unique resonance to the book. I am also especially grateful to the many artists who generously shared stories, photographs and unpublished plays with me for this book, who willingly answered my questions in interviews and by email and who sometimes read drafts of the sections of the book on their performances to check for accuracy or add special information: Tess Ellison and members of the community cast at Theatre by the Lake (*The Shepherd's Life*), Ffion Jones (*The Only Places We Ever Knew* and *Ode to Perdurance/Awdl Amser*), Ruth Levene and Anne-Marie Culhane (*A Field of Wheat*), Mike Pearson (*Carrlands*), Mary Swander (*Farmscape* and *Map of My Kingdom*), Octavio Solis (*Alicia's Miracle*) and Louise Anne Wilson (*The Gathering/Yr Helfa*). I want to thank Sarah Harper, artist and close friend, with whom I have collaborated numerous times, for always finding the 'wild card'—that unexpected 'bit' that makes her art so original. Thank you also to the Coalition of Immokalee Workers for their incredible work in transforming migrant farmscapes that I discuss in the book, but also for the time they have taken to give me interviews, find information and photographs, and answer questions, with special thanks to Greg Asbed, Marley Monacello, Julia Perkins and Nely Rodriguez. My heart-felt thanks to all the women in

UK agriculture who gave interviews and tours of their farms for *Who's Driving the Tractor?*, a PaR project I created with Sarah Harper, as part of the research for this book: Susannah Bolton (AHDB), Ali Capper (NFU and Stocks Farm), Rosemary Collier (Warwick Crop Centre), Caroline Drummond and Alice Midmer (LEAF), Emma Hamer (NFU and Meadowsweet Farm), Charlotte Hollins (Fordhall Farm), Sarah Pettitt and Freida Pettitt (Franklyn Farm), Marion Regan (Hugh Lowe Farms) and Becca Stevenson and Hannah Norman (Five-Acre Farm). I would like to thank my friend and colleague Baz Kershaw for our numerous conversations that challenged and inspired me and for our collaboration on *Prairie Meanders* in Iowa, an offshoot of his *Meadow Meanders*. A discussion of *Prairie Meanders* did not make it into the book, but its practice-as-research methodologies and thought experiments certainly did. I would also like give a special thank you to friends and colleagues at University of Warwick who encouraged me over the years of writing the book: Yvette Hutchison, who was always there for thrashing out ideas or just sharing laughter, Tim White who has been a great friend for well over a decade and has solved so many of my computer woes, Nicolas Whybrow who listened to my ideas and helped me secure internal funding for *Who's Driving the Tractor?*, Silvija Jestrovic and Bobby Smith for our numerous conversations that challenged and inspired me and who gave up precious time to read drafts of chapters, Nadine Holdsworth for her wise counsel and Rosemary Collier in Life Sciences who helped me understand the agricultural side of the story. I would also like to thank Deirdre Heddon and Sally Mackey, the editors of the series in which the book appears, for their strong support, excellent editorial advice, faith in my work and encouraging comments. I am so grateful for their insights and challenges that certainly improved the book.

On a more personal note, I want to acknowledge the amazing love and encouragement from my family without whom this book may never have been completed. My sons and their partners all supported the project in their special ways: discussing ideas and images, commenting on the artwork, sometimes even reading drafts and always offering reassurances that I would finish. I also want to thank David who has been by my side for many years and whose patience sustained me. My sister Sally and my dear friend and almost sister, Patricia, need special acknowledgements as they always stand by my side. And a very special 'thank you' needs to go to my first grandchild Torunn, who, although we lived eight time zones apart during the writing of the book, made sure that we 'played' almost every day once the pandemic changed our lives. Even though we could only meet on a screen, we cooked together, read, worked on art projects and celebrated 'dress up' days. And last but not least, I want to welcome the newest members to the family, Olivia, Margot, Alisair and Eleanor whose laughter and energy never cease to delight and inspire me.

Praise for *Performing Farmscapes*

"As agricultural subsidies disappear and pressures for adaptation in land usage mount, as food security and sustainability become key global issues, *Performing Farmscapes* is a timely and important reflection on human agency and its impacts in rural contexts, on contemporary resistances to its excesses, and on visions for recovering productive, agricultural landscapes—on the ground, in and through creative narratives and interventionist performances."

—Mike Pearson, *Professor Emeritus, Aberystwyth University, Honorary Professor, Exeter University*

"Fewer and fewer of us know how our food is produced, of the challenges farmers face, and about the impact of food production on nature. *Performing Farmscapes* creatively interprets the considerable value of performance as a means of raising our awareness of these issues and of highlighting the influence that we, as food citizens, can have through the choices we make."

—Professor Rosemary Collier *in Life Sciences, University of Warwick*

Contents

About the Author

Susan C. Haedicke is Emeritus Reader in Theatre and Performance Studies in the School of Creative Arts, Performance and Visual Cultures at University of Warwick in Coventry, England. She has published extensively on performances in public spaces and democratic participation in her book, *Contemporary Street Arts in Europe: Aesthetics and Politics* (Palgrave, 2013); several articles, including 'Breaking a Legacy of Hatred: Friches Théâtre Urbain's *Lieu Commun*' (2016) in *RIDE: The Journal of Applied Theatre and Performance* and 'Co-Performance of Bodies and Buildings: Compagnie Willi Dorner's *Bodies in Urban Spaces* and *Fitting* and Asphalt Piloten's *Around the Block*' (2015) in *Theatre Journal*; and book chapters, notably 'Street Arts, Radical Democratic Citizenship and a Grammar of Storytelling' in *The Grammar of Politics and Performance* (2015). Three essays written during this time looked at street performances that inserted farms into urban sites, 'Opéra Pagaï's *Entreprise de Détournement*: Collage of Geographic, Imaginary and Discursive Spaces' in *Theatre and the Politics of Space* (2012), 'Beyond Site-Specificity: Environmental Heterocosms on the Street' in *Performing Site-Specific Theatre* (2012) and 'Performing Farmscapes on Urban Streets' (2016) in *Popular Entertainment Studies*. In retrospect, it is clear that they signalled the beginning of a transition to her current area of research on performance and agriculture, and she has recently published several journal articles, including 'Coalition of Immokalee Workers: Farmworker-led Popular Education and Performance' (2020) in *RIDE: the Journal of Applied Theatre and Performance*, 'The *Aroma-Home* Community Garden Project's Democratic Narratives: Embodied Memory-Stories of Planting and Cooking' (2018) in *Public Art Dialogue* and '*Aroma-Home's* Edible Stories: An Urban Community Garden Performs' (2017) in *Renewable Agriculture and Food Systems*. She has also worked on practice-as-research projects on performance and food production, notably *Who's Driving the Tractor? Conversations with Women in UK Agriculture* (2018) and *Hope Is a Wooded Time* (2012–2014), both discussed in this book. In addition, she created *Prairie Meanders* in Iowa with Baz Kershaw in 2016 and *Grow Warwick* that imagined University of Warwick as an edible campus through performance installations set up in and around the Warwick Arts Centre in 2013. Now back in the United States, Haedicke plans to continue this research while also campaigning *for* Democratic candidates and *against* the current push for voting restrictions in many states, especially Florida where she now lives.

List of Figures

CHAPTER 1

Introduction: Farmscapes and Subjectivities

A Duet of Farmscapes and Performance

Imagine a farm … (Fig. 1.1).

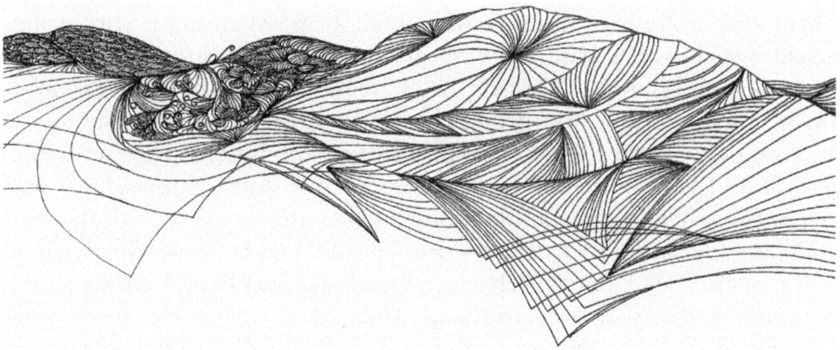

Fig. 1.1 *Hills and Valleys*. Pen and ink drawing by Kristen Oshyn

What do you see … and do? Are you walking in fields or pastures, gazing at them from a distance, planting and harvesting, driving a tractor, maybe? Can you smell the farm … hear it … feel it … taste it even?

Farmlands are inextricably linked to food production. Valued for their productivity, farms exist to ensure the survival of the earth's

© The Author(s), under exclusive license to Springer Nature Switzerland AG 2021

S. C. Haedicke, *Performing Farmscapes*, Performing Landscapes,
https://doi.org/10.1007/978-3-030-82434-1_1

growing populations. And yet, for many in the general public, farmlands are perceived as quintessential rural landscapes where pastures of grazing sheep and cows or fields of grain blowing in the wind are more picturesque than functional. Discarding this spurious dichotomy, *Performing Farmscapes* replaces the word *farmlands* with *farmscapes* to imply a hybrid of agricultural land that produces food and bucolic agrarian scenery that is 'visually' consumed and to shift the focus away from the *land* so prominent in the word *farmland*. In addition, I am using the word *farmscape* to acknowledge the recent 'performative turn' in landscape studies that privileges the lived experience of a landscape over a detached gaze (discussed later in the chapter) and, by implication, suggests that *farmscapes* live in the words, experiences and encounters of farmers, non-farmers and nonhumans; in the rhythms and repetitions of seasonal farming practices associated with a particular place; and in the traces and echoes left in the land by living creatures and natural forces alike. As part of the *Performing Landscapes* series, this book pairs farmscapes with performing and, in so doing, proposes a potential agency for the land to participate in the storytelling through its responses to human and nonhuman interventions, such as grassy mounds chronicling ancient burial sites, a wide valley traversed by a narrow stream recounting floods of long-ago or a forest's 'wolf tree' revealing its previous location in agricultural fields through the growth patterns of its branches (discussed in Chapter 3). *Performing Farmscapes* interrogates aesthetic, political and environmental implications of this duet of *farmscapes* and *performing* where human and nonhuman lives are entwined with the land as vibrant lifelines and pathways overlap and tangle to weave dynamic *storied* farmscapes.[1] James Rebanks, author and shepherd,[2] attests to the interconnectedness of lives and land when he describes the sheep farm landscape in the Lake District in England: '[t]he whole landscape here is a complex web of relationships between farms, flocks and families' (Rebanks 2016: 22). These relationships constitute embodied and visceral bonds between humans, nonhumans and a particular piece of land. For the sheep on the fells in the Lake District, this bond or sense of belonging is called 'hefting': a learned behaviour or 'homing instinct' of the sheep that is

[1] See Ingold (2011a: 141–175).

[2] Rebanks' autobiography, *A Shepherd's Life: A Tale of the Lake District* (2016), adapted by Chris Monks, is a case study in the next chapter. See also Rebanks (2020).

passed on by ewes to their lambs over many generations as the lambs graze with their mothers in a specific place on the fells each summer. The farmers feel that sense of belonging as well. George in Peter Gill's *The York Realist*, 'hefted' to his farm, knows he cannot follow his lover to London and leave the land without losing himself: 'I live here. I live here. You can't see that, though. You can't see it. This is where I live. Here!' (Gill 2001: 77). George's farm is his home, but more significantly, it is the cornerstone of who he is: this sense of identity rooted in the land is also inscribed in the term *farmscape*. Almost a half century ago, Yi-Fu Tuan gave this bond the name of 'topophilia': a 'human being's affective ties with the material environment' (Tuan 1974: 93), and his ideas still resonate today. He argues that the farmer's deep attachment to the land has entered his or her body: 'The entry of nature is no mere metaphor. Muscles and scars bear witness to the physical intimacy of the contact. The farmer's topophilia is compounded of this physical intimacy, of material dependence and the fact that the land is a repository of memory and sustains hope' (Tuan 1974: 97).

While the farmer may experience this bond to the land most acutely, farmscapes can kindle a strong attachment, even a sense of identity, in the non-farmer as well. As John Wylie claims in his accounts of his climb up Glastonbury Tor (Wylie 2002: 491–454) and his coastal walk on the South West Coast Path in England (Wylie 2005: 234–247), an experience of the countryside can confirm 'self-landscape relations' (Wylie 2005: 234), what he calls 'enlacements of self and landscape' (Wylie 2002: 442). Wylie asserts that embodied experiences in landscapes like these not only offer something to see but also awaken sensations, raise questions and spark imaginings that are not so much about the countryside, but more about the impact of the landscape on who he imagines himself to be. Inspired by Maurice Merleau-Ponty, he began to understand landscapes as visual and tangible encounters that gave rise to an emerging subjectivity that was coupled with the land. The pairing of farmscapes and performing helps articulate a farmscape's vibrant presence as an animated landscape that can contribute to an emerging subjectivity for those who experience it. The bond between landscape and subjectivity flourishes in storied farmscapes, and it can provide an opening for non-farmers to go beyond thinking about these agricultural landscapes as iconic rural scenery distinct from themselves and to explore how their encounters with farmscapes can contribute to their own emerging identities. Not surprisingly, many artists turn to performance to propel audiences into these affective farmscape experiences. In the performance-based case studies in *Performing Farmscapes*, the farmscape is viewed not as an object that needs to be explained but rather as vibrant matter that issues a call, that can make things happen—what Jane Bennett calls 'thing-power' (Bennett 2010: 1–19).

Performing Farmscapes is clearly not about agriculture, land use, food production or agri-food systems per se, and the book does not attempt to provide an overview of the extensive scientific research in the many varied topics that constitute agricultural studies. That does not mean however that debates about farming practices, food sovereignty,[3] exploitation of land and labour, global food chains and food democracy/food citizenship (discussed later in the chapter) have no place in the book. These important aspects of food production are inseparable from analyses of performance-based interventions into farmscapes. But neither is the book's focus on plays or performances that use farms as background scenery. Rather, it explores a duet of farmscapes and performing to understand how contemporary artists use performance and performativities not only to enable the public to see beyond the *objectness* of the farm landscape, but also to collaborate with farmscapes to remember past experiences, engage with present-day realities and realise future possibilities.

Performance and performativity play an increasingly important role in landscape studies as geographers shift from understanding landscape through perception to experiencing landscape through embodied encounters.[4] This phenomenological approach to landscape[5] was influenced by the ideas of Maurice Merleau-Ponty (*Phenomenology of Perception*, originally published in 1945) and Martin Heidegger ('Building Dwelling Thinking', originally published in 1954). Merleau-Ponty argues

[3] In The Nyéléni Food Sovereignty Forum Declaration, Mali (2007), the phrase 'food sovereignty' refers to the rights of peoples, communities and countries to determine agricultural (food production, distribution and consumption) policies that are socio-economically, culturally and environmentally appropriate to their local specificities (European Coordination Via Campesina: Nyéléni Europe, n.d.). See also Gioia (n.d.), La Via Campesina (n.d.), and World Development Movement (2012).

[4] See Wylie (2007) who traces the shift in landscape studies from a reliance on a 'disembodied gaze' (144) to 'the human subject's ongoing immersion in the world' (56).

[5] For overviews of landscape phenomenology, see Waterton (2013), Wattchow (2013), Wylie (2007: 139–186), and Wylie (2013).

that knowledge of the world is attained through the body, through everyday encounters. He does not diminish the value of 'seeing' the visible world, rather he inextricably links vision and the body and argues that embodiment is not just being *in* the world; it is being one with the world. The viewer and the landscape are intertwined through the gaze as the self performs the practices of the landscape and the landscape gives presence to the self.[6] Martin Heidegger introduces the notion of 'dwelling' as 'the manner in which mortals are on earth' (Heidegger 1993: 350). He explains that 'Man's relation to locales, and through locales to spaces, inheres in his dwelling. The relationship between man and space is none other than dwelling, thought essentially' (Heidegger 1993: 359). Dwelling is human essence inseparable from being alive; it is 'to be in the world ... as a homeland' (Young 2006: 373). Inspired by the notion of dwelling, Tim Ingold defines landscape as a 'meeting-up with' the land, as the world of dynamic and embodied encounters between humans and the land/landscape:

> A place owes its character to the experiences it affords to those who spend time there—to the sights, sounds and indeed smells that constitute its specific ambience. And these, in turn, depend on the kinds of activities in which the inhabitants engage. It is from this relational context of people's engagement with the world ... that each place draws its unique significance. (Ingold 2011b: 192)

A landscape is not solely an assemblage of topographical features or a blank slate on which humans and nonhumans write their stories, but rather a partner with whom humans and nonhumans interact, live and work. It is impossible to step out of the landscape or represent it objectively since we are part of that landscape and the landscape is part of us. As geographer Mitch Rose claims, 'the only thing that landscape ever *is* is the practices that make it relevant. While it appears as a definable material space, its materiality is constituted by the totality of possible performances immanent within it: the constitutive potential of the unfolding labyrinth' (Rose 2002: 462–463). These insights in landscape studies influenced my adoption of the word *farmscape* for this book. This integration of humans/nonhumans and the land, a merging of difference, reminds me of what Donna Haraway metaphorically calls 'the dance of world-making

[6] See also Merleau-Ponty (1971).

encounters' (Haraway 2008: 249) where, one could say, dissimilar dance partners find co-ordinated moves so that they can continue to dance even if they sometimes step on each other's toes. The relationship between dissimilar partners depends on mutual respect and a willingness to listen, trust, respond and change, a 'response-ability… that is also collective knowing and doing, an ecology of practices' (Haraway 2016: 34). In fact, the partners' very identities are defined by and depend on this partnership; they do not precede their interlacings. In this multispecies world, Haraway claims:

> Becoming-with, not becoming, is the name of the game; becoming-with is how partners are… rendered capable. Ontologically heterogeneous partners become who and what they are in relational material-semiotic worlding. Natures, cultures, subjects, and objects do not pre-exist their intertwined worldings. (Haraway 2016: 12–13)

For Haraway, the dancers are companion species, but bringing farmcapes into these worldings alongside humans and nonhumans is a key goal of this book.

Farmscapes acquire their unique character through the particular sensory experiences they offer to those who live there or who just pass through. As already discussed, the integration of landscape and identity is particularly powerful in farmscapes for both farmers and non-farmers. The emphasis on the body's sensuous engagement with farmscapes speaks to the concept of *affect* that assumes our ability both to be affected by a place, event or situation and to affect its direction or nature. Gilles Deleuze argues that affects 'are not feelings, they're becomings that spill over beyond whoever lives through them (thereby becoming someone else)' (Deleuze 1995: 136). For Deleuze, becoming is not a transformation of one thing into another, but a dynamic instability, a condition of chaos, difference and change, a way of asking and establishing new connections and new beginnings, often revealed through art. Focusing on the *affective/affected* body—its positioning, encounters, doings, reactions, emotions—signals a body in action, a body performing that, in turn, led to a 'performative turn' in landscape studies towards nonrepresentational (or more-than-representational) landscape theories.[7] In an early essay linking geography and performance, geographers Nigel Thrift and John-David Dewsbury suggest thinking about landscape as performed and practiced, as performing and practicing: 'a nonrepresentational outlook depends upon understanding and working with the everyday as a set of skills which are highly performative' (Thrift and Dewsbury 2000: 415).[8] Performance and performativity, they argue, encourage us to see place as 'lived' through daily practices, shared cultural artefacts and stories, and imagination (Thrift and Dewsbury 2000: 420–422).[9] David Crouch, too, insists on a 'rethinking of landscape as a process rather than an object; subjectively "in the making" rather as an assemblage of physical features' (Crouch 2013: 119). At the heart of this rethinking, he claims, are performance and performativity—what he calls 'the idea of *flirting* with space' (Crouch 2013: 119. Italics in original.) Crouch chose the term *flirting* for its sense of possibility where the 'unexpected opens out' (Crouch 2010: 1). These affective encounters with landscapes can last in body memories long after the person has left the place. Performance and performativity encourage a reassessment of landscape as 'moments of occurrence; things as they happen; conversations between things that happen' (Crouch 2013: 120).

An experience I had with a farm paralleled this trajectory in the development of landscape studies from perception to affective engagement. In 2017, I embarked on *Who's Driving the Tractor?*, a performance-as-research project on women in UK agriculture for which I was author and dramaturg. I will provide more details on the project in an extended case study in the next chapter, but for now all that is important to know is that my farmscape experience occurred when my collaborator and I went to interview Emma Hamer on her farm, Meadowsweet Farm, in Oxfordshire.

When we arrived at Meadowsweet Farm, a small family-run cattle and arable farm in Oxfordshire, England, we were greeted by an excited small black and white dog. Behind him at the door stood Emma Hamer. The first thing I noticed as I walked into her kitchen was a striking farm scene,

[7] See, for example, Anderson and Harrison (2010), Carolan (2008), Crouch (2010, 2013), Dewsbury (2000), Lorimer (2005), Olwig (2002, 2019), Thrift (2004, 2008), Waterton (2013), and Wattchow (2013).

[8] See also Dewsbury (2000: 473–496).

[9] See also Thrift (2004).

painted on the wall tiles behind the stove, with a farm house and barns, wheat fields, cows, sheep, chickens and a small dog. Emma began regaling us with stories about her farm work: caring for the cattle, starting a farm shop, her current profession as Senior Plant Health Advisor at the UK's National Farmers Union (NFU), and her research on neonicotinoids and insect health. Towards the end of the interview, I asked Emma about the scene on the painted tiles. 'It's right here, Meadowsweet Farm', she proudly exclaimed as she 'walked' us around the pristine painted farm pointing out the places we would see when we went outside. A misty rain started soon after we left the house. The little dog led the way past the barns and into the hilly pasture as the cows, to my great delight, ambled over to greet him by touching noses (Fig. 1.2).

Looking back, I am surprised how quickly I was drawn into the bucolic farmscape trope, but I was pulled out of that idealised reverie as that same cow came to greet me by pushing her wet nose against mine. I was no longer 'outside' the farmscape. My relationship shifted from the passive perception of a landscape painting to an embodied immersion in

Fig. 1.2 Meadowsweet Farm. 2017. Photographer: Susan Haedicke

the sounds and smells of the cows and in the feel of the soggy grass soaking the bottoms of my trousers and the damp wind sneaking down the back of my neck. I began to understand the farmscape very differently. This was not a representation that depicted a fixed and often romanticised version of a farm; this farmscape was a dynamic and unpredictable encounter that affected all my senses. The farm, for me, was no longer a collection of objects and physical features but asserted itself as a living process that invited me to dance in 'world-making encounters' (Haraway 2008: 249).

The case studies in *Performing Farmscapes* look at how the artists set up similar affective encounters so that audiences can *sense* the farmscape experience before seeking cognitive understanding. The book, in turn, tries to enable the reader to experience non-cognitive responses to farmscapes through Kristen Oshyn's drawings that accompany many of the chapters. While the drawings evoke the ideas explored in the chapter, they are not meant to illustrate the argument, but rather to enrich or complicate it through their ambiguities and playfulness. The abstract drawings are performative in J. L. Austin's sense of the word as a 'doing' that can make something happen (Austin 1976) as, I think, they draw the reader in without words in the way that farmscapes often do.[10]

Geographers clearly show their excitement at the prospect of using methodologies of performance studies to enliven geography by looking at landscape as a potentiality to experience rather than a fixed object to observe and analyse. And I, in turn, find inspiration from using insights from non-representational theories of landscape to engage with performance + farmscape projects where place, inhabitants (human and nonhuman) and passers-by are constantly in process of making and remaking themselves, their relationships and the farmscape; they are constantly in the process of 'performing farmscapes' as an 'event of place' (Massey 2005: 138–142). These performed and performing farmscapes are not mere backdrops or scenery for human endeavours, but rather, like the landscapes in modern drama that Elinor Fuchs analyses, perform alongside humans 'in a variety of roles, for instance, as mentor, obstacle, or ironist' (Fuchs 2002: 30). To perform a farmscape puts a spotlight on

[10] Kristen Oshyn grew up exploring a wooded acreage in rural Iowa and continues to be inspired by the natural world. She distils objects down to simple lines and then layers them to play with the intricacy born of simplicity. See her work on Instagram [oshyn.ink] and at Oshyn.Ink.

what the farmscape can do, proposes an exploration of agriculture's ongoing cultural and ecological processes as 'storied' lifeworlds, and explores farmscape as a presence that both binds living beings and the land and initiates its own affective networks.

When I ask farmers, scientists and social scientists about the role the arts can play in agricultural issues, they answer without hesitation *communication*. Many of the performances discussed in *Performing Farmscapes* do communicate scientific research about local and global issues affecting agriculture today to the general public in more accessible and engaging ways than an academic paper or a policy report, but it is also clear that the performances are not just agricultural information. The artists translate complex and abstract scientific data into absorbing life stories. They use creativity and embodied practices to develop research that can change the way people think about food systems, and they present that research in performances and performative events that can make the esoteric problems in the agri-food system comprehensible and compelling. This enacted research, immersed in the creative process and performance product, often reveals stories that people may not want to hear or creates encounters that contradict long-held beliefs or expose unexamined biases, and, in so doing, can encourage reflection not only on personal attitudes and assumptions, but also on what a fairer and more sustainable food system would look like. The most efficacious example of enacted research in this book is the protest art and activist performative events of the Coalition of Immokalee Workers, so this case study makes up the whole of Chapter 6. *Performing Farmscapes* interrogates how artists begin to address the general public's lack of knowledge about the agri-food system in order to embolden people to design alternatives to the ways we produce, distribute and consume food. It argues that their performances propose varied road maps that can animate spectators and participants to imagine and rehearse a new role in the food system: one that marks a shift from *food consumer* to *food citizen*.

Food Democracy/Food Citizenship

Yi-Fu Tuan, over forty years ago, observed that:

> [i]n modern life physical contact with one's natural environment is increasingly indirect and limited to special occasions. Apart from the dwindling

> farm population, technological man's involvement with nature is recreational rather than vocational. Sightseeing behind the tinted window of a coach severs man from nature. (Tuan 1974: 95–96)

His concerns have only increased in subsequent decades. As more than half of the world's population now lives in urban settings, fewer and fewer people have experienced farming or are even aware of the sources of their food. Where and how our foods are cultivated or raised and how they get from the farm to the supermarket shelves are details that are little known in many parts of the world as cooks are no longer restricted to seasonal and local produce and eaters often consume 'ready-made' meals with ingredients from farms only recognisable in the tiny print on the back of the package. Wendell Berry laments that lack of knowledge:

> Eating is an agricultural act. Eating ends the annual drama of the food economy that begins with planting and birth. Most eaters, however, are no longer aware that this is true. They think of food as an agricultural product perhaps, but they do not think of themselves as participants in agriculture. (Berry 2010 [1990]: 145)

Food production has become an abstract concept uncoupled from everyday experience and even from the natural environment. In the Introduction to a special issue on food labour of the journal *Organization*, the editors argue that as the agri-food system has become more industrialised and the world's population more urbanised, the processes that get food from the farm to the table have become hidden (Böhm et al. 2020: 195–212). Even the simplest of foods often relies on complicated supply chains that cross continents and so increase the distance food travels before it is consumed, creating what Tim Lang labels 'food-miles' (Lang 1999: 222). And the general public seems to have lost interest in this complex agri-food system. Historian Ann Vileisis explains to a non-specialised readership how we have become disconnected from food production and clarifies the damage that this lack of engagement and knowledge can wreak. The impact of this 'covenant of ignorance' (Vileisis 2008: 160) on today's food system and public health is significant, she argues, as we have relinquished public scrutiny over industrial producers and large grocery store chains and so can no longer ensure the availability of food that is responsibly and ethically grown, harvested and distributed.

Vileisis seeks to persuade consumers to shed this normalisation of ignorance and to become familiar and re-engaged with the story of our food (Vileisis 2008). Many scholars agree that engagement with food production, distribution and consumption can highlight the social and cultural significance of food, improve public health, discourage exploitation of land and labour, and ensure agricultural sustainability.[11]

Since the early 1990s, the concept of food democracy has gained traction as a way to address the public's disconnection from an increasingly industrialised and globalised agri-food system.[12] As this industrial food system reduced the agency of both farmers and consumers over food production, distribution and consumption and severed the population from farmlands, pressure began to build from below to regain access to participation in the process. Early advocates, Francis Moore Lappé (1990) and Tim Lang (1998, 1999), supported a democratisation of the agri-food system and an increase in citizen oversight over food chains in order to loosen corporate control. As Lang explains, 'it makes sense to see the dynamics of the food system as a titanic struggle between the forces of control and pressure to democratize' (Lang 1999: 218). Activist and scholar, Neva Hassanein provides an often quoted definition of food democracy:

> Food democracy seeks to expose and challenge the anti-democratic forces of control, and claims the rights and responsibilities of citizens to participate in decision-making. Food democracy ideally means that all members of an agro-food system have equal and effective opportunities for participation in shaping that system, as well as knowledge about the relevant alternative ways of designing and operating the system. (Hassanein 2003: 83)

[11] For problems caused by the disconnect between the public and the food system, see, for example, Coveney (2014), Kneafsey et al. (2008), Lang (1999, 2012), Lupton (1996), O'Kane (2016), Pretty (2002), and Shiva (2015).

[12] Scholarship on food democracy/food citizenship is vast. Civil Eats in the US and the UK Food Ethics Council offer extensive collections of recent statements, reports and essays on food citizenship (see especially Food Ethics Council 2015, 2017a, b; Cura and Crossley 2021). See also, Böhm et al. (2020), Gabriel and Lang (2015), Goldberg (2018), Hammelman et al. (2020), Hassanein (2003, 2008), Hatanaka (2020), Jhagroe (2019), Kneafsey et al. (2008), Knezevic et al. (2017), Lang (1998, 1999, 2012), Lappé (1990), Levkoe (2011), and Lozano-Cabedo and Gómez-Benito (2017).

In a later essay, Hassanein identifies five key dimensions of food democracy: (1) effective coalitions working towards sustainability, (2) knowledge about food and food systems, (3) ongoing discussions and idea-sharing around food issues, (4) an understanding of how to improve the food system and (5) an orientation towards the community good (Hassanein 2008: 290–291).

Key to food democracy is public participation in the food system, so to construct a food democracy, uninformed and disengaged *food consumers* must become informed and engaged *food citizens* who participate in food behaviours that can change the food system locally, regionally, nationally and globally. The food citizen, therefore, must be knowledgeable, pro-active in decision-making and governance, more interested in the common good than individual gain, and willing to adopt a practice of 'commoning' where resources, ideas, tasks and benefits are shared.[13] The Food Ethics Council lays out a way to accomplish this transformation.[14] The story of the current food system is one of 'what the consumer wants', but what if we change that story, the writers ask. Consumerism is a 'mindset' of ideas that prioritises selfishness and a reluctance to care for others (human and nonhuman) and the environment, but when that mindset changes, the system changes. With a food citizen mindset, people want to shape their food system rather than just act within pre-determined limits, and they no longer see food as a commodity of personal choice to consume, but rather they understand that their food choices have a meaningful impact on the food system as a whole. Thinking of the public as a community of food citizens can encourage changes in the agricultural community as well. 'Food Citizenship: How thinking of ourselves differently can change the future of our food system' reimagines the roles of all participating in the food system (shareholders, producers, retailers, government, NGOs and a citizen public) through a lens of the citizen mindset to explore how working together can transform the food system (Food Ethics Council 2017b: 19–31).[15] Public response to these

[13] See Elinor Ostrum, *Governing the Commons: The Evolution of Collective Institutions for Collective Action* (1990). She won the Nobel Prize in Economics for her work on the value of the commons (also called common property resource or common pool resource) in 2009.

[14] See Food Ethics Council (2017a, b).

[15] See also Food Ethics Council (2015, 2021) and Cura and Crossley (2021) for more details on food citizenship.

ideas has been an exponential increase in alternative food initiatives, such as CSAs (community-supported agriculture), allotments and communal gardens, slow food movements, Fairtrade campaigns and more, that give people more control over food policies, what is grown and how it is grown (Levkoe 2011; Hammelman et al. 2020). These alternative food initiatives engage the public directly in the agri-food system, but the arts can play an important role in encouraging food citizenship as well. A crucial way to change a *consumer* mindset into a *citizen* mindset is through language. Anna Cura and Dan Crossley, citing 'Food Citizenship: A Communications Toolkit', argue that 'exposure to the word "consumer" significantly decreases the public's sense of responsibility in shaping the world around us' (Cura and Crossley 2021: 8) whereas the word *citizen* empowers people to care and to participate. 'As citizens, we care about animals being treated humanely, about the wellbeing of the environment, about the livelihood of those who grow and make our food. … Words lead to stories. Stories told many times create new mindsets' (Cura and Crossley 2021: 8). Stories provide more than information; they can change the way people think, feel and respond. Artists are particularly adept at crafting compelling narratives in words and images, so their contributions to promoting food citizenship can be significant.

Performing Farmscapes argues that the artists in the case studies use stories, performances and performative practices to facilitate a transformation from food consumer to food citizen in their works even if they do not mention food citizenship or food democracy by name and, in so doing, not only inform and engage the public in agriculture, but also challenge unexamined assumptions and long-held beliefs through storytelling and motivate consumer activism by politicising food. Some of the performance-based work also creates spaces to experiment with and 'rehearse' new ideas and behaviours through discussions, workshops and collaborative art-making. This creative approach can break down resistance to change both within and outside the farming community and encourage often radical thinking about what an improved agri-food system could be. Mary Swander's plays, for example, often verbatim pieces relying on extensive interviews with farmers, clearly address pressing agricultural issues, like foreclosure, the loss of family farms to corporations or farmland transfer, but she also ends each performance with a conversation between the audience and invited experts or farmers about how individual farmers and engaged non-farmers can collaboratively move forward on issues raised in the plays (Chapter 2). Charlotte Hollins leads an informative and performative walk around her family's farm, Fordhall Organic Community Farm, that narrates food citizenship through the story of how she and her brother saved their tenanted farm from developers through a crowd-sourcing initiative and, in so doing, did not sell shares solely as a financial transaction but also as a way to invite the public to have a say in what they wanted the farm to be (Chapter 2). The cost of the membership fee to join the Collective works in a similar way and its dialogic practices of food citizenship enable participants to become decision-makers on the participating farm in Ruth Levene and Anne-Marie Culhane's *A Field of Wheat* (Chapter 4). The coalition-building among farmworkers, students, religious groups and food citizens in the Coalition of Immokalee Workers and its farmworker-led activism and art demonstrate food democracy in action (Chapter 6). The latter two case studies offer particularly innovative and effective ways to construct food democracy and so receive longer, in-depth analyses.

A research methodology often associated with food citizenship is participatory action research (PAR), an umbrella term for a range of methodological investigations that seek to engage communities in collaborative problem-solving, social action and knowledge construction to achieve greater social justice. PAR draws inspiration from the radical pedagogy of Paulo Freire (see Chapter 6). John Gaventa, a pioneer in PAR and former Director of the Highlander Research and Education Center[16] in Tennessee renowned for its work in social justice, offered a definition decades ago that is still accurate:

> Participatory [action] research attempts to break down the distinction between the researchers and the researched, the subjects and objects of knowledge production by the participation of the people-for-themselves in the process of gaining and creating knowledge. In the process, research is seen not only as a process of creating knowledge, but simultaneously, as education and development of consciousness, and of mobilization for action. (Gaventa 1988: 19)

[16] The main office building of Highlander Center holding decades of archives was burned to the ground in April 2019. Right-wing extremists are the suspected arsonists in part because of the discovery of a symbol associated with white supremacy spray-painted in the car park.

PAR, often associated with initiatives seeking to construct food democracy and food citizenship, gives tools to communities to understand the constructed-ness of their particular situation and to find solutions to transform their environment. Their specific experiences can then be applied to a larger surrounding context. Gaventa, now at the Citizenship Development Research Centre at University of Sussex, in collaboration with Gregory Barrett analyses the efficacy of active citizenship, encouraged in PAR projects, to solve issues of injustice and oppression: 'citizen engagement makes a difference to achieving both material and democratic goals' (Gaventa and Barrett 2012: 2399) because 'an informed and aware citizenry … can participate in democratic life, hold the state to account, and exercise their rights and responsibilities effectively' (Gaventa and Barrett 2012: 2402).[17] Participatory action research is often associated with community projects around a range of social issues, frequently in agriculture,[18] but it has been adapted by many artists working in applied theatre with specific communities as well. The case studies in *Performing Farmscapes*, but particularly in Chapters 4 and 6, explore how artists are adapting PAR dialogic practices in their performance-based projects to empower the public to become informed and engaged food citizens.

Before turning to the specific case studies, I will introduce a theoretical framework that underpins the analyses of the case studies. I propose three key distinct (although simultaneously overlapping) modalities of farmscapes that link ways of knowing and experiencing them and that contribute to understanding how performances encourage participation in practicing food democracy. The first modality, local farmscapes, focuses on the significance of the local for an understanding of farmscapes; the second, working farmscapes, on the importance of recognising farmscapes as working landscapes; and the third, naturalcultural[19] farmscapes, on the inseparability of nature and culture in storied farmscapes. While in this chapter, I explore these three modalities of farmscapes separately, it is important to note that the subsequent chapters interweave them within the analyses of the case studies. Some discussions of specific performances or performative events may highlight one of the modalities over another,

[17] See also Ansley and Gaventa (1997) and Gaventa and McGee (2010).

[18] See, for example, Schattman et al. (2015).

[19] This term is from Haraway (2016).

but all three modalities co-exist in every farmscape and play important roles in interpretations of the farmscape performances.

Home-Grown: Local Farmscapes

Agricultural landscapes exist all over the world, but each farmscape is a local place, home-grown with its own life-stories and place-stories. Brian Wattchow who writes extensively on landscape and culture admits that he finds it:

> impossible to write about landscape and place only in the abstract. For me, landscapes and places are best considered from the perspective of the particular. Every place is a result of an ongoing interaction between natural and cultural phenomena. Human expectations and desires for a location and the resulting way that humans live in a landscape shape and are shaped by that location. The reciprocity between people and locations on Earth's surface provides the reference point for all considerations of landscapes and places. (Wattchow 2013: 87)[20]

As far back as the second century, Claudius Ptolemy recognised the value of appreciating the local and introduced the concept of chorography as a way 'to understand and represent the unique character of individual places' (Cosgrove 2004: 59), primarily through the skills of storytellers and artists to depict these specificities. In *Performing Farmscapes*, both Ffion Jones' *The Only Places We Ever Knew* (Chapter 2) and Mike Pearson's *Carrlands* (Chapter 3) draw on insider knowledge to exemplify the power of the storyteller to illuminate the particularity of place. In Jones' guided walk around her family farm, her father regaled the audience with anecdotes from family history and information about sheep farming in Wales. Pearson's *Carrlands*, on the other hand, offers self-guided audio-walks that link history, legend, topography and ecology in three locations in north Lincolnshire, England.

Lucy Lippard argues that inhabitants' narratives of place are certainly written *about* the landscape, but they also written *into* it and help to construct it. Like Wattchow, she sees 'landscape' not as an abstract term, but as an intersection of everything there that touches its inhabitants: what she chooses to call *place*. The 'lure of the local' is:

[20] See also Legat (2008).

> the pull of place that operates on each of us, exposing our politics and our spiritual legacies. Inherent in the local is the concept of place—a portion of land/town/cityscape seen from the inside, the resonance of a specific location that is known and familiar... entwined with personal memory, known or unknown histories, marks made in the land that provoke and evoke. Place is latitudinal and longitudinal within the map of a person's life. It is temporal and spatial, personal and political. A layered location replete with human histories and memories... . It is about connections, what surrounds it, what formed it, what happened there, what will happen there. (Lippard 1997: 7)

For Lippard, a *sense of place* 'does indeed emerge from the senses. The land, and even the spirit of the place, can be experienced kinetically, or kinesthetically' (Lippard 1997: 34). Lippard's concept of place parallels many geographers notion of landscape, and it certainly reflects my notion of *farmscapes*. Peter Howard claims, 'Landscape... is always about 'my place', or at least somebody's place' (Howard 2011: 2). For him, landscape is a visceral response to the place rather than the land itself, it is experiences and encounters occurring in that place over time. In *Carrlands*, Pearson describes the significance of the local for the residents:

> Don't ever think of this place as essentially empty. For a handful at least, it is a lived landscape, a place of memory and emotion, of fine-grained knowledge, of drives and compulsions and pleasure and anxieties where the land styled the lives, just as the lives styled the land, where you live in hope that what you got back would repay what you put in, where you could let the farm get the boss of you if you weren't careful. ... These people are insiders, the ones who would never need to ask for directions. (*Carrlands*, 'Hibaldstow Carrs 4: Earth')

This inseparability of place, time and identity is key to an understanding of local farmscapes that act as constellations of place-time narratives where place is less spatial than temporal as it occurs in an encounter of humans, nonhumans, ideas, memories, objects. It is important not to romanticise the local however. Doreen Massey warns that the notion of *place*, often evoking references to 'local place':

> can have negative resonances as often as positive ones as it has come to have totemic resonance. For some it is the sphere of the everyday, of real

> and valued practices, the geographic source of meaning, vital to hold on to as 'the global' spins its ever more powerful and alienating webs. For others, a 'retreat to place' represents a protective pulling up of drawbridges and a building of walls against the new invasions. (Massey 2005: 5)

And other scholars identify the conflation of the local with goodness, community and well-being as the 'local trap' that elides injustices and exploitations.[21]

The farmscape performed and performing in each of the case studies responds to a specific place that resonates with a past and summons a future. Wattchow maintains that the local landscape is where we '*experience* the "invisible threads" of the relationships that come together to constitute the phenomenon that *is* place' (Wattchow 2013: 93). While that is obvious in the case of the site-specific work, it is also important in plays and performance installations that tour the world, like Agnes Denes' *Wheatfield: a Confrontation* and Shelley Sacks' *Exchange Values* (Chapter 4) or Cherríe Moraga's *Heroes and Saints* (Chapter 5).[22] While many of the issues, debates and even farming practices are global (in part as a result of colonisation), the particularities of the local make up the farmscape narratives. In another part of the globe, the tropes would be different.

Working Farmscapes

Agriculture is a narrative of progress achieved through hard work. Geographer Don Mitchell looks at how working agricultural landscapes like large industrial farms are produced, how the farmlands in a capitalist economic system are the end product of human labour that transforms raw materials into commodities that can be sold on the local and global market for profit. Mitchell argues that it is essential to recognise that these working landscapes produce inequalities and injustices in capitalist systems

[21] See Gray (2014) and Guthman (2014).

[22] Critical place inquiry from social sciences is 'research that takes up critical questions ... informed by the embeddedness of social life in and with places.... An increased focus on place in critical research matters because it enables greater attention to the ways in which the land and environmental issues intersect with social issues and social life' (Tuck and McKenzie 2015: 2). An adaptation of this approach recurs throughout the book as performances are interrogated, in large part, through the social and agricultural life in and with the places they perform.

(Mitchell 1994, 2003). Working landscapes like farmlands, Wylie explains in his discussion of Mitchell's ideas, must be:

> understood in terms of the axioms of classical Marxism: the labour theory of value, commodity fetishism, the need for surplus value, or profit, ongoing struggles between the interests of capital and labour. … In order to fully understand the forms of land use and occupancy we see when we gaze upon any urban or agricultural scene we have to refer to systems of capitalist production and reproduction. (Wylie 2007: 106)

While an in-depth analysis of Marx's theoretical apparatus is beyond the scope of this book, his concepts of the social relations of production and commodity fetishism are important to an understanding of why so many of the case studies seek to dispel the perception of farmlands as bucolic rural landscapes. For Marx, the organisation of work and the experience of labour are disguised in a capitalist mode of production. These social relations of production determine the relationship between the capitalists (owners of the means of production) and the workers (who do not own the means of production). Since capitalism, based on private ownership and profit, depends on achieving the greatest productive capacity at the least cost, the social relations of production between owner and worker inevitably create antagonisms and exploitation. Many of the case studies in Chapters 5 and 6 reveal how in capitalism, the goal is to extract as much labour as possible from the workers while paying subsistence wages in order to acquire the greatest profit for the owners. The 1960 documentary, *Harvest of Shame* (Chapter 5), exposed the exploitative working conditions of migrant farmworkers living in sub-standard housing and on poverty wages that made milk for their children too expensive. Edward R. Murrow, who narrated the documentary, points out the repetitive, tedious and gruelling nature of the work as the workers followed the harvests year after year and highlights their powerlessness in this controlled social and economic relationship as they admitted they could see no way out of the situation. In a different way, Octavio Solis, in his unpublished play, *Alicia's Miracle*, performed in 2015, draws attention to the capitalist push towards realisation of the fullest production capacity by revealing a disregard for the workers' health in the quest for greater crop yields. Dramatised from a report completed by the Center for Investigative Reporting in 2014[23] on the use of toxic pesticides in the California strawberry fields, Solis personifies the three opposing positions in this controversy—the 'owners' (chemical companies or large-scale farmers) seeking the largest profit possible, the scientists searching for accurate data on risks and rewards, and the workers whose bodies live the results of these decisions—in the confrontation of the three main characters (Chapter 5).

For Marx, the capitalist relations of production are generative of workers' embodied experiences, and the latter cannot be understood without analysis of the former. In the case of agriculture, the products of the workers' labour, agricultural commodities, erase the labour. The commodities are not only the crops and livestock however. Mitchell points out how an agrarian landscape itself 'is much like a commodity: it actively hides (or fetishizes) the labour that goes into its making… those who study landscape representations are repeatedly struck by how effectively they erase or neutralise images of work' (Mitchell 1998: 103). The 'agricultural landscape', writes Mitchell, 'is a place of work, of toil, of intensive, difficult labor' (Mitchell 2003: 234). Commodity fetishism analyses the way misunderstandings of labour and value are systematically produced in capitalist economies, and it highlights the systematic misunderstanding of the *social* character of labour, *social* in this context referring to relationships between participants in the production process. Marx argues that when a thing produced by human labour becomes a commodity through its participation in capitalist market exchange, the labour that went into its construction becomes invisible. The commodity appears to have intrinsic value of its own, but actually its value comes from the human toil that produced it. The invisibility of that labour, however, renders the work done by the workers value-less. One can see immediately the relevance of this analysis for interpretations of farmscapes. If the value resides in the commodity—the fruit, vegetables, grains, meat—and if the work on the farm is invisible, it is easy to imagine these sites as idyllic pastoral farmscapes, free from the inequalities and injustices of capitalist systems. Thus it is easy *not* to see the labour of the farmworkers. Drawing on Marx, Mitchell labels agricultural products 'dead labor' since the 'living labor', the actual work done by the farmworkers, disappears into the product itself (Mitchell 2003: 233–248).

[23] Yeung et al. (2014).

Fig. 1.3 *A PerFarmance Project West Midlands*. 2016. Five-Acre Farm, Ryton-on-Dunsmore, England. Performer: Juan Manuel Aldape. Photographer: Susan Haedicke

In Juan Manuel Aldape and Chris Bell's *A PerFarmance Project West Midlands*,[24] performed on Five-Acre Farm, a CSA (community-supported agriculture) farm in Ryton-on-Dunsmore in England in 2016, one of the early scenes exposed the concept of 'dead labor' (Mitchell 2003) as it encouraged the audience to 'see', and even experience, the labour of farming. At the start of the performance, the audience was divided into two groups. I joined the group that was directed to walk across the large muddy field ready for planting to a distant corner where we saw a wheelbarrow piled high with soil and an actor with a teaspoon. The actor was taking soil from the wheelbarrow to a specific place in the field several steps away where he would bend over and carefully tip the soil from the spoon to create a new pile (Fig. 1.3).

[24] See *PerFarmance Project*, https://perfarmanceproject.wordpress.com/ and *PerFarmance Project West Midlands* 2019, https://blogs.warwick.ac.uk/perfarmance/ for additional information and photographs.

The work was slow and tedious: back and forth, back and forth as the task was repeated. We watched for several minutes wondering what else would happen. Music that had been playing in the background began to get louder and faster, and as the music's tempo increased so did the speed of the farmworker's activity. Still we just watched as he began to sweat and breathe heavily. In spite of his many trips between the wheelbarrow and the pile, the wheelbarrow was still full and the pile still very small. Some spectators looked perplexed, some even started to edge away, but the actor carried on with his task. Finally, one spectator joined the activity and began to carry handfuls of soil to the pile. Then another joined and another until we were all moving the soil. This simple performative moment not only transformed an act of spectatorship into a 'doing', but more importantly it focused attention on the labour to produce the product—the pile of soil in the field. That simple task could stand in for any number of tasks needed to prepare a field for planting crops that would become the farm products, but the theatrical moment shifted all the attention from the product to the work. We had to *see* the toil, first with our eyes and then with our bodies. The 'dead labor' pile disappeared into the 'living labor' (Mitchell 2003)—the actual work we did. As we worked, we could see the other half of the audience way across the field walking blind-folded between two rows of lettuces—they literally could not see the labour needed to produce those crops they brushed past.

The working farmscape is what keeps the farm sustainable. For farms, sustainability is not only environmental, but also economic and social. So the working farmscape is certainly the labour in the soil or with the livestock, but it also includes the work involved in farm management, book-keeping, marketing, and community and industry outreach. Without this additional labour (labour often done by women), the farm would not be viable, so it is crucial to think beyond working the land in working farmscapes. This modality of working farmscapes is part of the analysis of all the performing farmscapes in the book, so I will cover it in more depth in subsequent chapters, but particularly in Chapter 6, where I look at the social relations of production alongside commodity fetishism enacted in the protest art of the Coalition of Immokalee Workers.

Nature, Culture and the 'Call to Care': Naturalcultural Farmscapes[25]

Farmlands are often identified as 'cultural landscapes'.[26] Carl Sauer is credited with coining the phrase when, in 1925, he wrote in 'The Morphology of Landscape' that 'the cultural landscape is fashioned from a natural landscape by a culture group. Culture is the agent, the natural area is the medium, the cultural landscape is the result' (Sauer 2007: 63). Sauer sought an understanding of the *culture* that made the landscape, and that culture, he claims, could be discovered in 'place facts' or things found in the landscape. He argues that human cultures mould a landscape's form,[27] a premise that implies a passive natural world acted upon or 'tamed' by human intervention. His ideas grew out of a belief that culture and nature were two separate domains, with culture in the dominant position, and he and other geographers sought to understand or imagine the relation between them.

The concept of 'cultural landscape' has evolved significantly since 1925. Clifford Geertz explains:

> Both [culture and nature] initially separate the works of man and the processes of nature into different spheres — 'culture' and 'environment' — and then attempt subsequently to see how as independent wholes these externally related spheres affect one another. With such a formulation, one can ask only the grossest of questions: 'how far is culture influenced by the environment?' 'How far is the environment modified by the activities of men?' And can only give the grossest of answers: 'To a degree, but not completely'. (Geertz 1963: 2)

Geertz seeks a more integrated interpretation of nature and culture rather than a hierarchical one implied by Sauer.[28] Wylie explains that until

[25] The phrase 'call to care' is from Rose (2006: 542). The term 'naturalcultural' is from Haraway (2016).

[26] See Roe and Taylor, *New Cultural Landscapes* (2014).

[27] See Sauer (1956).

[28] See also Whatmore (1999) where she explains 'hybrid geographies' that 're-cognize the "human" in "human geography"' and asserts the importance of reintegrating the nonhuman into the human world; Descola (2013), for whom the cultivated fields and pastures of farmlands are key sites where humans and nonhumans interact: 'theatres of subtle sociability' (Descola 2013: 5) that are 'beyond nature and culture' as the title of his

relatively recently, '[t]hinking of nature and culture—as distinct and independent realms' (Wylie 2007: 10) was the norm, but this separation, he argues, raises:

> political, moral and ethical dilemmas. Were humans once a part of nature? If so, how and when did they separate themselves off from it? If so, does this mean that some human creatures are more 'natural' than others? Is nature then fixed and given and culture dynamic and pliable? Or are cultural practices simply responses to natural environmental conditions? (Wylie 2007: 10)

For Mitch Rose, 'the cultural landscape is not an object that reflects "culture" nor a milieu where culture takes place but something that happens as we actively mark the world to orient our unfolding selves' (Rose 2006: 539). He claims that cultural landscapes are 'possibilities' that provoke, question, challenge: not destinations, but inclinations, and he suggests an approach to exploring cultural landscapes not as relationships between nature and culture, but as 'dreams of presence' that are 'impossible possibilities. They are things that do not exist and are, strictly speaking, impossible. And yet, they exist as possibilities, as things held out to us as possible at the horizon of our being' (Rose 2006: 542). Dreams of presence are quests to understand our identity, our place within the world, our very existence, and they are illusive attempts to find certitude, to hold on to stability. As such, they are inseparable from what Rose identifies as a 'call to care' (Rose 2006: 542): 'an inclination to gather, associate, and (yes) to take comfort. It is a call to invest in, cultivate, and nurture' (Rose 2006: 543). Rose argues that human existence is grounded in care for ourselves, others, both human and nonhuman, and for the world we inhabit. As such, this 'call to care' involves some form of action—and interaction—and thus has the potential to effect social and political change.[29] For Rose, culture, as expressed in our stories, objects, customs or beliefs, is not the result of care, but rather signifies a 'moving toward' horizons of understanding and affect that can never

book attests; and Cloke and Jones who argue 'place characteristics are often wrapped up in the dynamic presences and absences of living beings which are commonly constructed, corporately, as nature' (2004: 313), and so it is essential to take nonhuman agency seriously. See also the edited collection by Harrison et al. (2004).

[29] See also Kneafsey et al. (2008).

be reached—a 'becoming'. The cultural landscape is less a reflection of the relationship between nature and culture than an engagement with an affective response to the world around us. That response, in turn, contributes to a sense of self, as Wylie found on his walks up Glastonbury Tor and along the South West Coast Path, discussed earlier in the chapter. Rose cites Deleuze and Guattari who argue that sensory stimuli from moving through the world have a significant impact on human subjectivity, and he uses that claim to assert that this bond between self and environment plays a key role in eliminating the culture/nature binary and, ultimately, in promoting care for our world. 'This is landscape that is not being constructed but engaged with' (Rose 2006: 549). Being attuned to the stories in the landscape, therefore, means that 'it is being heard. It is a landscape that speaks to the people, beckoning them to care' (Rose 2006: 549).

Haraway, too, discards the nature-culture binary with her merging of the two words into *natureculture*. She begins by assuming that nature and culture are what they are because of their interlaced 'naturalcultural' worldings and that a naturalcultural approach advocates a prominence of care. In her discussions on 'becoming-with' companion species, she introduces the notion of 'response-ability', a term that implies a 'call to care', a responsibility towards another, a need to listen, look back and respect another. Early in *When Species Meet*, she narrates Barbara Smuts' steep learning curve as she figured out how to research baboons. Smuts gradually understood that her neutral, objective stance seemed a threat and that she had to 'become-with' the baboons, to participate in a two-way responsive relationship. In that way, she showed and received respect and engaged in a kind of dialogue (Haraway 2008: 23–25).[30] Haraway later argues that these *becoming-with* worldings and *response-abilities* need to become viral to 'infect processes and practices that might yet ignite epidemics of multispecies recuperation' (Haraway 2016: 114): mutual respect, response and care across species. While Haraway focuses on cross-species relationships, I think she opens a door to 'becoming-with' land/landscapes in her narrative of the symbiotic relationship between ants and acacias, a mutualism of protection and nurture (Haraway 2016: 122–125) and her exploration of sheep's knowledge of the land on which they graze (Haraway 2016: 96).

[30] See also Haraway (2016). I am particularly fond of her discussion of pigeons, pp. 14–29.

These ideas are particularly relevant to farmscapes as neglect, or even denials, of the inseparability of nature and culture, self and environment, lead to exploitation of the land and labour in the insatiable push for cheap and abundant food for the consumer and large profits for the grower. Many of the case studies in *Performing Farmscapes* issue a 'call to care', but it can be seen most blatantly in the socially-engaged performance practices discussed in Chapter 4 and the political and protest art around migrant farmscapes in Chapters 5 and 6.

My Farmscapes? Beginnings of an Answer

Where do I fit into a study of farmscapes as I did not grow up on a farm and do not farm now? My connection to the performance-based farmscape case studies in *Performing Farmscapes* is clearly rooted in my training as a theatre and performance studies scholar and teacher and my artistic practice as a dramaturg/writer. As I reflect back on my decades of teaching, I find an abundance of farm plays on my syllabi: Susan Glaspell's *Trifles* (1916), John McGrath's *The Chevriot, the Stag and the Black Black Oil* (1973), Sam Shepard's *Buried Child* (1978), Caryl Churchill's *Fen* (1983), Sue Glover's *The Bondagers* (1991), Pearl Cleage's *Flyin' West* (1995), David Harrower's *Knives in Hens* (1996), Rita Dove's *The Darker Face of the Earth* (1999), Richard Bean's *Harvest* (2005), Marina Carr's *On Raftery's Hill* (2009), Brendan Jacobs-Jenkins' *Appropriate* (2014) and DC Moore's *Common* (2017) are just a few that dramatised farmscapes and vividly established the bond between the land and the farm family. Working at University of Warwick has provided me with opportunities to create or collaborate on performance-based projects that explore various aspects of farmscapes and agriculture. The first was *The Tocil Wood Project* (2008) in which the students, a professional artist and I created a site-specific performance intervention depicting local history stories in an ancient woodlands on the Warwick campus. But this woodland project inadvertently drew me into agricultural farmscapes as the wood, surrounded by farmlands, was a rich source of farmland legends of centuries-old agricultural practices, family feuds and even murder. *Hope is a Wooded Time* (2012–2014), a durational community project sited in what used to be part of the vast peach orchards in Montreuil, just outside Paris, France, is discussed in Chapter 4. Other projects included *Grow Warwick* (2013), a series of performance installations that imagined University of Warwick as an edible campus; *Aroma Home* (2013–2014), a community garden project in Villetaneuse, a banlieue north of Paris, created with the site's neighbours, mostly recent migrants from Africa and the Near East[31]; and *A PerFarmance Project West Midlands* (2016), briefly discussed earlier in this chapter. *Prairie Meanders* (2016) in Grinnell, Iowa, that I co-designed with Baz Kershaw, created two prairie walks that enabled the participants to observe the entire world on a human scale as the maze-like pathways simulated the earth's ocean and wind currents, but they also learned about local environmental and agricultural events in small installations placed along the walks, for example, the number of chickens culled in the 2015 outbreak of avian flu in several midwestern states in the United States or the amount of wheat produced in various locations around the world. In addition, these paths tangentially explored the tensions between prairie and agricultural lands in Iowa. The most recent project, *Who's Driving the Tractor? Conversations with Women in UK Agriculture* (2018), a case study in Chapter 2, receives extended analysis as it looks at local versions of agrarian feminism and gendered farmscapes in the UK. These performance-as-research projects where performance is not only an end in itself, but acts as a key research methodology enabled me to experiment with varied scholar positionings as I developed the projects: participant-observer, 'life historian', auto-ethnographer, research scholar and artist. I will discuss these scholarly positionings in more detail as I move into the various case studies.

While my teaching and artistic practices played a significant role in the writing of this book, I came to farmscape stories much earlier as I grew up surrounded by family legends about ancestors who, in the late seventeenth century, sailed from England to try their luck on rocky Concord, Massachusetts farms—land the family owned for centuries and where I often played as a child. I remember a woodcut, now lost, of my great-great-great-grandfather repairing a stone wall on his Concord farm, and I spent many hours trying to find its traces on the land. Farm legends also abounded about later generations of the family who packed all their belongings in a covered wagon to join the movement westward to find more fertile farmlands. And I came to farmscapes through embodied encounters. On my bookshelf at home, I have a small faded photograph of me at about four years old surrounded by piles of vegetables I had picked in my grandparents' large kitchen garden as I happily eat a raw ear

[31] See Haedicke (2017, 2018).

of corn. As I grew older, I found myself drawn to living history farms and farm museums. And I did not hesitate when, a few years ago, a Wisconsin farmer asked me if I would help harvest the wheat by driving the combine.

But what kind of a book have I written as a response to these varied agrarian encounters? The performance-based projects on which I worked clarified for me the importance of embodied farmscape experiences and research that demanded a live presence—a 'being-there'. This emphasis on fieldwork as a key methodological frame certainly limited the geographic scope of the book, but it also highlighted the significance of a kind of insider knowledge—not as a farmer necessarily, but as a participant-observer with knowledge of the socio-cultural context of the local specificities. As a consequence the performing and performed farmscapes discussed in the case studies are located primarily in the United States where I grew up and the United Kingdom where I have lived for close to two decades. I was able to attend most of the contemporary works I discuss, but for those I did not see, I relied on historical ethnography and archival material and/or conducted extensive interviews and email exchanges with the artists who often shared creative process documents with photographs, mind maps and storyboards as well as video documentation of the performance, if available. In addition, if the performance had been site-specific, I visited the location and walked its route.

Yet as I began to reflect on these projects, I recognised that I was particularly drawn to their often overt activism, their compulsion to address pressing issues in agriculture and difficulties for farming families and to encourage their audiences to care and even act. The works I gravitated towards foregrounded knowledge-exchange and political engagement. I realised that many of these case studies applied theoretical and practical ideas proposed in analyses of food citizenship to their creative processes. The audience-participants were placed in a role similar to that of food citizens as they shifted from consumers of the art to engaged 'citizens' and decision-makers. Two case studies in particular, Ruth Levene and Anne-Marie Culhane's year-long participatory project *A Field of Wheat* (Chapter 4) and the Coalition of Immokalee Workers' varied forms of performative protest art (Chapter 6), provide innovative and complex examples of this approach, so they receive extended thick description and analysis of their creative use of concepts and methodologies associated with food democracy's imperative of public engagement and decision-making in the agri-food system. While I was most interested in the political nature of the case studies' performances, I also found myself drawn to the beauty of agricultural landscapes. I did not want to dismiss this aesthetic joy in the stunning vistas or somatic delight in encounters with farm animals, but I wanted to dispel the sense that farmlands were bucolic agrarian scenery devoid of hard work. In an extended analysis of Louise Ann Wilson's *The Gathering/Yr Helfa* (Chapter 3), a site-specific walking performance on a sheep farm on the slopes of Mount Snowdon in Wales, I unpack this tension between iconic rural vistas and working farmscapes and explore the role the beauty plays.

The book focuses on a limited number of case studies to explore ways performed and performing farmscapes can participate in debates on the socio-political and economic working landscapes of agriculture to promote social justice and make progress towards environmental sustainability. Several of these case studies could be explored in more than one chapter. I tried to choose case studies that both clearly foregrounded local specificities and effectively used their insider knowledge of practices and issues to enter into and particularise dialogue. It is important to note, however, that the local issues they raise extend well beyond their specific locale and are vigorously debated on the global stage as well. In making these choices, many excellent examples were left out.

The Structure of *Performing Farmscapes*

One of the most difficult challenges in developing the structure of *Performing Farmscapes* came from an initial attempt to use categories of various types of farms (e.g. family farms vs. industrial farms, or farms based on specific crops or livestock they produce) as an organisational principle. Early on, I recognised the need to limit the geographical area since trying to cover the diversity of farmscapes worldwide would have resulted in a rather superficial encyclopaedic approach. As I started my research looking at examples of performed and performing farmscapes in the British and American context, it at first seemed obvious that I should contrast small family farms and large-scale 'industrial' farms. They seemed to fit neatly into that structure as the profound links between the farming family and the land are explored in many plays and site-specific works whereas intensive industrial farming practices with their excessive use of fertilisers and pesticides and exploitation of land and labour are the subjects of political and protest art. But as I looked more closely, I realised that division in agriculture was anything but neat. Many family farms are actually quite large, use fertilisers and pesticides extensively, and concentrate on just one crop or farm animal almost exclusively. Many so-called industrial farms are family-owned, farmed only by family members, or grow crops organically. The use of migrant farmworkers, especially at harvest, tends to be determined by crop rather than by farm ownership or size, and environmental sustainability and ideals of food justice can be found on a wide range of farms just as labour exploitation, lack of biodiversity and soil depletion exist on varied farms across the agricultural sector. So-called family farms[32] have a particular presence in the public imaginary that suggests an incomplete or misleading agricultural image of pastoral beauty and peace, and by extension, fairness and justice. The celebratory rhetoric of this 'agrarian imaginary' or 'agrarian ideal' of 'family farming' implies a place of unity, nurturing, harmony and working together under equitable conditions—all characteristics of the cultural ideal of 'family'. In this telling, these farms are like a family, caring for and cherishing all the human and nonhuman members. But sociologists, notably Julie Guthman (2004, revised in 2014), Stephanie Lewthwaite (2007) and Margaret Gray (2014),[33] have challenged that 'agrarian imaginary' of the family farm as a proxy for social justice and exposed the ways in which using the metaphor of 'family' conceals patterns of exploitation and abuse both within the family itself and towards farmworkers and the land. Guthman argues that organic farming in California, rather than 'meeting the ideal of "farming in nature's image"' (Guthman 2014: 26), replicates the structures and abuse of large commercial agriculture that seeks to maximise productivity and profitability. The agrarian myth, argues Lewthwaite, with its emphasis on pastoral beauty of farm landscapes obscures the labour of the farmworkers who create them. Gray concurs as she looks at the invisibility and consequent exploitation of migrant farmworkers on family farms, small and large, in New York.

Artists, too, have worked to dispel the agrarian imaginary. Certainly, playwrights have explored tensions between the love of the land and the toil it demands. This hard labour, hidden in farm landscape scenery, has appeared in plays such as Caryl Churchill's *Fen* (1983), Susan Glover's *The Bondagers* (1991) and Rita Dove's *Darker Face of the Earth* (1996,

[32] See Taber (2019) who argues that family farming has never worked and that today the emotional and financial hardships of solo farming can affect quality of life and result in abuse. Instead farming must become collaborative ventures.

[33] See also Mitchell (2003).

first stage production), to name just a few. Other plays look at larger socio-political issues that have an impact on farms, like Bea Roberts' *And Then Come the Nightjars* (2015) that dramatises one farm's plight during the UK 2001 Foot and Mouth epidemic, John Somers' devised community performance, *The Living at Hurford* (2002), that looks at the aftermath of Foot and Mouth on rural communities or Annabel Souter's *Seeds* (2013) that dramatises an actual court case in Canada between a farmer and the huge biotech corporation Monsanto over genetically modified seeds. Playwrights have also tackled assumptions of justice, equality and care promoted by the agrarian ideal. Plays like Marina Carr's *On Raftery's Hill* (2000), Pearl Cleage's *Flyin' West* (1995), Susan Glaspell's *Trifles* (1916) and Sam Shepard's *Buried Child* (1978) are notable examples that expose the intra-family abuse, injustices and neglect, but certainly not the only ones. Rather than spend time on these plays that have often received detailed scholarly analyses, I wanted to look at less familiar performances that sought to understand farmscapes as lived experiences from the inside to reveal the shrouded work on the farm. As a result, I have relied on the ideas and creative strategies used by the artists to determine the book's structure: farm stories told by the farmers or by visible topographical features in the land, socially-engaged projects, political plays, documentaries and performative protest art. The rest of the book argues that the work of the artists in these case studies plays a significant role in promoting food citizenship (even though the artists do not use that term) and in providing a space to rehearse those behaviours. It explores how these works contribute to reconnecting not only farming and non-farming communities, but also living creatures and our *natural-cultural* world, and it asks why that matters. Haraway describes Ursula Leguin's 'carrier bag theory of fiction' that suggests parallels between storytelling and 'capacious bags for collecting, carrying and telling the stuff of life' (Haraway 2016: 118). These metaphoric sacks hold seemingly unrelated objects, images, ideas, memories that, like a collage, come together in creative practice. I have tried to create such a capacious bag of performed and performing farmscapes in this book that looks at works ranging from plays and theatrical performances, both in theatres and on site, to durational community (or applied) performance projects and performative political events like hunger strikes, protests or marches. Some of the case studies of performed and performing farmscapes provide opportunities to farmers to tell their own stories; some look to the past in order to look forward; others facilitate difficult conversations about contentious local and global socio-political, economic and environmental issues facing agriculture today; some expose exploitations and abuses in farmscapes to encourage audiences to join the fight for social and food justice; and still others displace the focus on the human 'actor' and instead centre on the agency of the farmscape itself.

Chapters 2 and 3 look at how the artists encourage food citizenship by informing and engaging audiences in farming through the three modalities of local, working and naturalcultural farmscapes. Chapter 2, 'Performing Storied Farmscapes', interrogates how performance and performative events narrate the complexities of farmscapes through the words of farmers and in the repetitions of their farming tasks. The chapter looks at dramatised autobiography in James Rebanks and Chris Monks' *The Shepherd's Life*, variations of verbatim performance in Theatre Passe Muraille's *The Farm Show* and Mary Swander's *Farmscape* and *Map of My Kingdom*, storied walks around farms in Ffion Jones' *The Only Places We Ever Knew*, Charlotte Hollins' guided walk on Fordhall Farm and Sylvia Grace Borda's *Farm Tableaux Finland*, and a performance-as-research project, *Who's Driving the Tractor?*, that gave women in UK agriculture an opportunity to share their challenges and achievements in a performance-based form. Chapter 3, 'Conversations with Farmscapes: Traces and Echoes', looks at performances that rely on farmscapes to narrate their agrarian stories through their language of agricultural remains, topographical features and traces left on the land. It interrogates how Ffion Jones' *Ode to Perdurance/Awdl Amser*, Friches Théâtre Urbain's *Hope is a Wooded Time*, Louise Ann Wilson's *The Gathering/Yr Helfa* and Mike Person's *Carrlands* make those traces visible through very different creative practices. Each one, however, attempts to diminish the nature/culture binary and encourage care for the environment by revealing farmscape stories told by and on the land. Chapter 4, 'Cultivating Dialogues Beyond the Farm', looks at socially-engaged, performative and participatory projects in which farmscapes acquire a presence beyond the actual farm through encounters between farmlands (actual, virtual, 're-sited' and globalised) and engaged participants around a range of social and (agri)cultural issues. Agnes Denes' *Wheatfield: A Confrontation*, Nessie Reid's *The Milking Parlour*, Shelley Sacks' *Exchange Values*, and Ruth Levene and Ann-Marie Culhane's *A Field of Wheat* promote food citizenship by providing knowledge about the farmscapes they are performing, but also by creating spaces for dialogue and decision-making for the participants on how to improve food production processes and providing opportunities to establish coalitions. Chapters 5 and 6 continue an emphasis on participatory activity and look at the role of political activism in the performing of farmscapes as they narrow the focus to a specific farmscape in the American agricultural landscape: the migrant farmscape. The exploitation of migrant farmworkers is certainly not unique to the American context, but rather is a global human rights issue. And, in fact, the Coalition of Immokalee Workers (Chapter 6) shares its strategies, both political and artistic, with other human rights organisations fighting for justice for workers around the world. Looking at global migrant farmworker injustice and arts activism could fill another book, so I limited my chapters to the United States. Chapter 5, 'Exposing Hidden Migrant Farmscapes', focuses on political and protest art that seeks to make visible the deplorable working conditions of migrant farm workers in the United States as it examines various ways artists expose the exploitation of these hidden workers. Starting with the legacy of John Steinbeck's *The Grapes of Wrath*, published in 1939, it then looks at the 1960 documentary, *Harvest of Shame*, on migrant farmworkers; the early protest art and *actos* of El Teatro Campesino in partnership with the farmworkers' union, United Farm Workers; and two plays, Octavio Solis' *Alicia's Miracle* and Cherríe Moraga's *Heroes and Saints*, that expose the dangerous health risks of extensive pesticide use on California farms. Chapter 6, 'Transforming Migrant Farmscapes: Performative Protest Art of the Coalition of Immokalee Workers', offers an in-depth analysis of the role of protest art and performative political events developed by the human rights organisation, the Coalition of Immokalee Workers, in transforming the migrant farmscape through human rights activism and protest art. This extended case study looks at the efficacy of their campaigns for justice for farmworkers through Worker-led Social Responsibility and by developing committed food citizens. Chapter 7, 'Conclusion: Sowing Seeds and Cultivating Conversations for Future Harvests', reviews the book's core argument about the role performance and performative events can play in introducing and promoting food democracy and food citizenship, and it places the urgent issues explored through the performances in a larger context food insecurity in the global agri-food system. The book emphasises a diversity of performance-based activities and events rather than proposing a specific genre of performance practice that can be themed as 'farmscape performance'.

In this book, I seek to offer thick descriptions and critical reflections on the storied farmscapes and embodied encounters that inform, engage and affect the audiences within and outside farming communities and to understand how the artists encourage conversations, coalitions and activism around pressing global issues in food production within a local context. I look at how the *liveness* of performance can begin to lessen the communication gap between farmer and consumer/citizen, issue a 'call to care', encourage 'response-ability' and imagine a more just and sustainable agricultural future. Here, I hope to shift the focus away from the object-ness of both the farmscape and the art and instead to explore how the narratives, drawings and performances suggest tangled paths and impossible possibilities towards a food democracy.

Works Cited

Ansley, Fran and John Gaventa. 1997. 'Researching for Democracy and Democratizing Research'. *Change: The Magazine of Higher Learning* 29.1: 46–53.

Anderson, Ben and Paul Harrison. 2010. 'The Promise of Non-Representational Theories'. In *Taking-Place: Non-Representational Theories and Geography*. Eds. Ben Anderson and Paul Harrison. Farnham, Surrey and Burlington, VT: Ashgate. 1–35.

Austin, J.L. 1976 [1955]. *How to Do Things with Words: the William James Lectures Delivered at Harvard University in 1955*. 2nd ed. Oxford: Oxford University Press.

Bean, Richard. 2005. *Harvest*. London: Oberon Books Ltd.

Bennett, Jane. 2010. *Vibrant Matter: A Political Ecology of Things*. Durham and London: Duke University Press.

Berry, Wendell. 2010 [1990]. *What are People for? Essays*. 2nd ed. Berkeley, CA: Counterpoint.

Böhm, Steffen, Marja Spierenburg, and Tim Lang. 2020. 'Fruits of Our Labour: Work and Organisation in the Global Food System'. *Organisation* 27.2: 195–212.

Carolan, Michael S. 2008. 'More-than-Representational Knowledge/s of the Countryside: How We Think as Bodies'. *Sociologia Ruralis* 48.4: 408–422.

Carr, Marina. 2009. *Plays 2: On Raftery's Hill; Ariel; Woman and Scarecrow; The Cordelia Dream; Marble*. London: Faber and Faber.

Churchill, Caryl. 1983. *Fen: Drama*. London and Toronto: Samuel French, Inc.

Civil Eats. 2021. https://civileats.com/.

Cleage, Pearl. 1999. *Flyin' West and Other Plays.* New York: Theatre Communications Group.

Cloke, Paul and Owain Jones. 2004. 'Turning in the Graveyard: Trees and the Hybrid Geographies of Dwelling, Monitoring and Resistance in a Bristol Cemetery'. *Cultural Geographies* 11.3: 313–341.

Cosgrove, Denis. 2004. 'Landscape and Landschaft'. *GHI Bulletin No. 35*: 57–71.
Council of Europe (COE). 2000. European Landscape Convention and Reference Documents. https://rm.coe.int/CoERMPublicCommonSearchServices/DisplayDCTMContent?documentId=09000016802f80c6.
Coveney, J. 2014. *Food*. Abingdon, Oxon: Routledge.
Crouch, David. 2010. *Flirting with Space: Journeys and Creativity*. Farnham, UK and Burlington, VT: Ashgate.
Crouch, David. 2013.'Landscape, Performance and Performativity'. In *The Routledge Companion to Landscape Studies*. Eds. Peter Howard, Ian Thompson and Emma Waterton. London and New York: Routledge. 119–127.
Cura, Anna and Dan Crossley. 2021. 'Harnessing the Power of Food Citizenship'. Food Ethics Council. https://foodcitizenship.info//wp-content/uploads/2019/10/fec_food_citizenship_report_final.pdf.
Deleuze, Gilles. 1995. *Negotiations: 1972–90*. Trans. Martin Joughin. New York: Columbia University Press.
Descola, Philippe. 2013. *Beyond Nature and Culture*. Trans. Janet Lloyd. Chicago and London: The University of Chicago Press.
Dewsbury, John-David. 2000. 'Performativity and the Event: Enacting a Philosophy of Difference'. *Environment and Planning D: Society and Space* 18: 473–496.
Dove, Rita. 1999. *The Darker Face of the Earth*. London: Oberon Books.
European Coordination Via Campesina: Nyéléni Europe. n.d. https://www.eurovia.org/campaign/nyeleni/.
Food Ethics Council. 2015. 'This is the #citizenshift: A Guide to Understanding and Embracing the Emerging Era of the Citizen'. https://drive.google.com/file/d/1ce2qAxXjgw30KuuZTH9BKTw1xCUGV7Us/view.
Food Ethics Council. 2017a. 'Food Citizenship: A Communications Toolkit'. https://www.foodethicscouncil.org/resource/food-citizenship-a-communications-toolkit/.
Food Ethics Council. 2017b. 'Food Citizenship: How Thinking About Ourselves Differently Can Change the Food System'. https://www.foodethicscouncil.org/app/uploads/FoodCitizenship2017_web.pdf.
Food Ethics Council. 2021. https://www.foodethicscouncil.org/.
Fuchs, Elinor. 2002. 'Reading for Landscape: The Case of American Drama'. In *Land/Scape/Theatre*. Eds. Elinor Fuchs and Una Chaudhuri. Ann Arbor: University of Michigan Press. 30–50.
Gabriel, Yiannis and Tim Lang. 2015 [1995]. *The Unmanageable Consumer*. 3rd ed. Los Angeles: Sage.
Gaventa, John. 1988. 'Participatory Research in North America'. *Convergence: An International Journal of Adult Education*. 24.2–3: 19–48.

Gaventa, John and Rosemary McGee. 2010. 'Introduction: Making Change Happen—Citizen Action and National Policy Reform: Making Change Happen'. In *Citizen Action and National Policy Reform: Making Change Happen*. Eds. John Gaventa and Rosemary McGee. London and New York: Zed Books. 1–43.

Gaventa, John and Gregory Barrett. 2012. 'Mapping the Outcomes of Citizen Engagement'. *World Development* 40.12: 2399–2410.

Geertz, Clifford. 1963. *Agricultural Involution: The Process of Ecological Change in Indonesia*. Berkeley: University of California Press.

Gill, Peter. 2001. *The York Realist*. London: Faber and Faber, Ltd.

Gioia, Paula. n.d. European Food Sovereignty Movement. http://orfc.org.uk/wp-content/uploads/2017/01/The-European-Food-Sovereignty-Movement-Paula-Gioia.pdf.

Glaspell, Susan. 1987. *Trifles*. In *Plays by Susan Glaspell*. Ed. C.W.E. Bigsby. Cambridge: Cambridge University Press. 35–46.

Glover, Sue. 2005. *Bondagers & Straw Chair*. London: Methuen.

Goldberg, Ray A. 2018. *Food Citizenship: Food Systems Advocates in an Era of Distrust*. Oxford: Oxford University Press.

Gómez-Benito, Cristóbal and Carmen Lozano. 2014. 'Constructing Food Citizenship: Theoretical Premises and Social Practices'. *Italian Sociological Review* 4.2: 135–156.

Gray, Margaret. 2014. *Labor and the Locavore: The Making of a Comprehensive Food Ethic*. Berkeley and Los Angeles: University of California Press.

Guthman, Julie. 2014 [2004]. *Agrarian Dreams: The Paradox of Organic Farming in California*. 2nd ed. Oakland, CA: University of California Press.

Haedicke, Susan. 2017. '*Aroma-Home*'s Edible Stories: An Urban Community Garden Performs'. *Renewable Agriculture and Food Systems* 32.4: 1–6.

Haedicke, Susan. 2018. 'The *Aroma-Home* Community Garden Project's Democratic Narratives: Embodied Memory-Stories of Planting and Cooking'. *Public Art Dialogue* 8.1: 114–130.

Hammelman, Colleen, Charles Z. Levkoe, Julian Agyeman, Sanjay Kharod, Ana Moragues Faus, Elisa Munoz, Jose Oliva, and Amanda Wilson. 2019/2020. 'Integrated Food Systems Governance: Scaling Equitable and Transformative Food Initiatives through Scholar-Activist Engagement'. *Journal of Agriculture, Food Systems and Community Development* 9.2: 71–86.

Haraway, Donna. 2008. *When Species Meet*. Minneapolis, MN: University of Minnesota Press.

Haraway, Donna. 2016. *Staying with the Trouble (Experimental Futures): Making Kin in the Chthulucene*. Durham, NC: Duke University Press.

Harrison, Stephan, Steve Pile and Nigel Thrift, eds. 2004. *Patterned Ground: Entanglements of Nature and Culture*. London: Reaktion Books, Ltd.

Harrower, David. 2015. *Knives in Hens*. London and New York: Bloomsbury Methuen Drama.

Hassanein, Neva. 2003. 'Practicing Food Democracy: a Pragmatic Politics of Transformation'. *Journal of Rural Studies* 19: 77–86.

Hassanein, Neva. 2008. 'Locating Food Democracy: Theoretical and Practical Ingredients'. *Journal of Hunger and Environmental Nutrition* 3.2–3: 286–308.

Hatanaka, Maki. 2020. 'Beyond Consuming Ethically? Food Citizens, Governance, and Sustainability'. *Journal of Rural Studies* 77: 55–62.

Heidegger, Martin. 1993 [1954]. 'Building Dwelling Thinking'. In *Basic Writings*. Ed. David Farrell Krell. London: Routledge, 343–263.

Howard, Peter. 2011. *An Introduction to Landscape*. Farnham, UK: Ashgate.

Ingold, Tim. 2011a. *Being Alive: Essays on Movement, Knowledge, and Description*. Abingdon, UK: Taylor & Francis Group.

Ingold, Tim. 2011b [2000]. *The Perception of the Environment: Essays on Livelihood, Dwelling and Skill*. 2nd ed. London and New York: Routledge.

Jacobs-Jenkins, Branden. 2019. *Appropriate*. London: Nick Hern Books.

Jhagroe, Shivant. 2019. 'Food Citizenship and Governmentality: Neo-Communitarian Food Governance in The Hague'. *Politics and Governance* 7.4: 190–201.

Kneafsey, Moya, Rosie Cox, Lewis Holloway, Elizabeth Dowler, Laura Venn, Helena Tuomainen. 2008. *Reconnecting Consumers, Producers and Food: Exploring Alternatives*. Oxford: Berg Publishers.

Knezevic, Irene, Alison Blay-Palmer, Charles Z. Levkoe, Phil Mount, Erin Nelson, eds. 2017. *Nourishing Communities: From Fractured Food Systems to Transformative Pathways*. Cham, Switzerland: Springer International Publishing AG.

La Via Campesina: International Peasants Movement. n.d. https://viacampesina.org/en/.

Lang, Tim. 1998. 'Towards a Food Democracy'. In *Consuming Passions: Food in the Age of Anxiety*. Eds. Sian Griffiths and Jennifer Wallace. Manchester: University of Manchester Press. 13–24.

Lang, Tim. 1999. 'Food Policy for the 21st Century: Can It Be Both Radical and Reasonable?' In *For Hunger-proof Cities: Sustainable Urban Food Systems*. Eds. Mustafa Koc, Rod MacRae, Luc J.A. Mougeot and Jennifer Welsh. Ottawa, ON, Canada: International Development Research Centre. 216–224.

Lang, Tim. 2012. 'Food Industrialisation and Food Power: Implications for Food Governance'. In *Taking Food Public: Redefining Foodways in a Changing World*. New York and London: Routledge. 11–22.

Lappé, Francis Moore. 1990. 'Food, Farming, and Democracy'. In *Our Sustainable Table*. Ed. Robert Clark. San Francisco: North Point Press. 143–159.

Legat, Allice. 2008. 'Walking Stories; Leaving Footprints'. In *Ways of Walking: Ethnography and Practice on Foot*. Eds. Tim Ingold and Jo Lee Vergunst. Farnham, Surrey, England and Burlington, VT: Ashgate Publishing Ltd. 35–49.

Levkoe, Charles Zalman. 2011. 'Towards Transformative Food Politics'. *Local Environment* 16.7: 687–705.

Lewthwaite, Stephanie. 2007. 'Race, Paternalism, and "California Pastoral": Rural Rehabilitation and Mexican Labor in Greater Los Angeles'. *Agricultural History* 81.1: 1–35.

Lippard, Lucy R. 1997. *The Lure of the Local: Senses of Place in a Multicentered Society*. New York and London: The New Press.

Lorimer, Hayden. 2005. 'Cultural Geography: The Busy-Ness of Being "More-Than-Representational"'. *Progress in Human Geography* 29.1: 83–94.

Lorzano-Cabedo, Carmen and Cristóbal Gómez-Benito. 2017. 'A Theoretical Model of Food Citizenship for Analysis of Social Praxis'. *Journal of Agricultural and Environmental Ethics* 30: 1–22.

Lupton, D. 1996. *Food, the Body and the Self*. London: Sage Publications Ltd.

Massey, Doreen. 2005. *For Space*. Los Angeles, London, New Delhi, Singapore, Washington, DC: Sage.

McGrath, John. 1981. *The Chevriot, the Stag and the Black, Black Oil*. London: Methuen.

Merleau-Ponty, Maurice. 1971 [1947]. 'Cezanne's Doubt'. In *Sense and Nonsense*. Trans. Hubert L. Dreyfus and Patricia Allen Dreyfus. Evanston, IL: Northwestern University Press. 9–25.

Merleau-Ponty, Maurice. 2012 [1945]. *Phenomenology of Perception*. Trans. Donald A. Landes. London and New York: Routledge.

Mitchell, Don. 1994. 'Landscape and Surplus Value: the Making of the Ordinary in Brentwood, CA'. *Environment and Planning D: Society and Space* 12: 7–30.

Mitchell, Don. 1998. *Cultural Geography: A Critical Introduction*. Oxford: Basil Blackwell.

Mitchell, Don. 2003. 'Dead Labor and the Political Economy of Landscape: California Living, California Dying'. In *The Handbook of Cultural Geography*. Eds. Kay Anderson, Mona Domosh, Steve Pile and Nigel Thrift. London: Sage. 233–248.

Moore, D.C. 2017. *Common*. London and New York: Bloomsbury Methuen Drama.

Nebbe, Charity and Ben Stanton. 3 May 2018. 'AgArts Promotes Intersection of Agriculture and Art'. Interview with Mary Swander. Iowa Public Radio. https://www.iowapublicradio.org/post/agarts-promotes-intersection-agriculture-and-art#stream/0.

O'Kane, Gabrielle. 2016. 'A Moveable Feast: Contemporary Relational Food Cultures Emerging from Local Food Networks'. *Appetite* 105: 218–231.

Olwig, Kenneth R. 2002. *Landscape, Nature, and the Body Politic: From Britain's Renaissance to America's New World*. Madison, WI: University of Wisconsin Press.

Olwig, Kenneth R. 2019. *The Meanings of Landscape: Essays on Place, Space, Environment and Justice*. London and New York: Routledge.

Oshyn, Kristen. 2021. *Oshyn Ink*. https://oshyn.ink/.

Ostrum, Elinor. 1990. *Governing the Commons: The Evolution of Collective Institutions for Collective Action*. Cambridge: Cambridge University Press.

Pearson, 2007. *Carrlands*. http://www.carrlands.org.uk/.

PerFarmance Project. https://perfarmanceproject.wordpress.com/.

PerFarmance Project West Midlands, 2016. https://blogs.warwick.ac.uk/perfarmance/.

Pretty, Jules. 2002. *Agri-Culture: Reconnecting People, Land and Nature*. London: Earthscan.

Rebanks, James. 2016. *The Shepherd's Life: A Tale of the Lake District*. UK and USA: Penguin Random House.

Rebanks, James. 2020. *English Pastoral: an Inheritance*. UK and US: Allen Lane/Penguin Books.

Robinson, Jo. 2016. *Theatre and the Rural*. Basingstoke: Palgrave Macmillan.

Roberts, Bea. 2015. *And Then Come the Nightjars*. Nick Hern Books.

Roe, Maggie and Ken Taylor, eds. 2014. *New Cultural Landscapes*. Abingdon, Oxon: Routledge.

Rose, Mitch. 2002. 'Landscape and Labyrinths'. *Geoforum* 33.4: 455–467.

Rose, Mitch. 2006. 'Gathering "Dreams of Presence": A Project for the Cultural Landscape'. *Environment and Planning D: Society and Space* 24: 537–554.

Sauer, Carl O. 1956. 'The Agency of Man on Earth'. In *Man's Role in Changing the Face of the Earth*. Ed. W. Thomas. Chicago, IL: University of Chicago Press.

Sauer, Carl O. 2007 [1925]. 'The Morphology of Landscape'. Originally published in *Geography* 2.2: 19–54. Repr. In John A. Wiens, Michael R. Moss, Monica G. Turner and David J. Mladenoff. *Foundation Papers in Landscape Ecology*. NY: Columbia University Press. 36–70.

Schattman, Rachel E., Hannah M. Aitken, V. Ernsto Méndez and Martha Casswell. 2015. 'Climate Change Resilience on Vermont Farms: A Research Report for Service Providers'. *ARLG Research Brief*, no. 2. Agroecology and Rural Livelihoods Group. University of Vermont.

Shepard, Sam. 1980. *Buried Child*. In *Buried Child, Seduced, Suicide in B: Plays*. London: Faber and Faber.

Shiva, Vandana. 2015. *Who Really Feeds the World?* London: Zed Books.

Solis, Octavio. 2015. *Alicia's Miracle*. Unpublished play manuscript.

Soutar, Annabel. 2012. *Seeds*. Vancouver, BC, Canada: Talonbooks.

Taber, Sarah. 16 June 2019. 'America Loves the Idea of Family Farms. That's Unfortunate'. *The Intelligencer*. https://nymag.com/intelligencer/amp/2019/06/america-loves-the-idea-of-family-farms-thats-unfortunate.html.

Thrift, Nigel. 2004. 'Performance and Performativity: A Geography of Unknown Lands'. In *A Companion to Cultural Geography*. Eds. James S. Duncan, Nuala C. Johnson, and Richard H. Schein. Malden, MA, Oxford, UK, and Carlton, Australia: Blackwell. 121–136.

Thrift, Nigel. 2008. *Non-Representational Theory: Spaces, Politics, Affect*. London and New York: Routledge.

Thrift, Nigel and John-David Dewsbury. 2000. 'Dead Geographies—And How to Make Them Live'. *Environment and Planning D: Society and Space* 18: 411–432.

Tuan, Yi-Fu. 1974. *Topophilia: A Study of Environmental Perception, Attitudes, and Values*. New York: Columbia University Press.

Tuck, Eve and Marcia McKenzie. 2015. *Place in Research: Theory, Methodology, and Methods*. New York and London: Routledge.

Vileisis, Ann. 2008. *Kitchen Literacy: How We Lost Knowledge of Where Our Food Comes From and Why We Need to Get It Back*. Washington, Covelo and London: Island Press/Shearwater Books.

Waterton, Emma. 2013. 'Landscape and Non-Representational Theories'. In *The Routledge Companion to Landscape Studies*. Eds. Peter Howard, Ian Thompson, and Emma Waterton. London and New York: Routledge. 66–75.

Wattchow, Brian. 2013. 'Landscape and a Sense of Place: a Creative Tension'. In *The Routledge Companion to Landscape Studies*. Eds. Peter Howard, Ian Thompson, and Emma Waterton. London and New York: Routledge. 66–75.

'What Is Food Ethics? A Short Explainer by the Food Ethics Council'. 2018. https://vimeo.com/272330915.

Whatmore, Sarah. 1999. 'Hybrid Geographies: Rethinking the "Human" in Human Geography'. In *Human Geography Today*. Eds. Doreen Massey, John Allen, and Philip Sarre. Cambridge, UK: Polity Press. 87–96.

World Development Movement. 2012. 'Food Sovereignty'. https://www.globaljustice.org.uk/sites/default/files/files/resources/food_sov_tricky_questions.pdf.

Wylie, John. 2002. 'An Essay on Ascending Glastonbury Tor'. *Geoforum* 33: 441–454.

Wylie, John. 2005. 'A Single Day's Walk: Narrating Self and Landscape on the South West Coast Path'. *Transactions of the Institute of British Geographers* 30.2: 234–247.

Wylie, John. 2007. *Landscape*. Abingdon, Oxon and New York, NY: Routledge.

Wylie, John. 2013. 'Landscape and Phenomenology'. In *The Routledge Companion to Landscape Studies*. Eds. Peter Howard, Ian Thompson, and Emma Waterton. London and New York: Routledge. 54–65.

Yeung, Bernice, Kendall Taggart and Andrew Donohue. 2014. 'California's Strawberry Industry is hooked on Dangerous Pesticides'. *Reveal*, from the Center for Investigative Journalism. https://www.revealnews.org/article/californias-strawberry-industry-is-hooked-on-dangerous-pesticides/.

Young, Julian. 2006. 'The Fourfold'. In *The Cambridge Companion to Heidegger*. 2nd ed. Ed. Charles B. Guignon. Cambridge: Cambridge University Press. 373–392.

CHAPTER 2

Performing Storied Farmscapes

A farmscape is a storied world, an unfinished multispecies time-place narrative told through words, practices and topographical features. A 'storied world', writes Tim Ingold, is:

> a world of movement and becoming, in which any thing—caught at a particular place and moment—enfolds within its constitution the history of relations that have brought it there. In such a world, we can understand the nature of things only by attending to their relations, or in other words, by telling their stories. For the things of this world *are* their stories. (Ingold 2011a: 160. Italics in original.)

Ingold argues that storytelling is a key way to enter into a landscape, to experience it from within. He dismisses the idea that a story is a 'tapestry' to cover the world or to 'clothe it with meaning' (Ingold 2011b: 56). Rather, stories 'serve to conduct the attention of performers *into* the world, deeper and deeper, as one proceeds from outward appearances to an ever more intense poetic involvement. At its most intense, the boundaries between person and place, or between the self and the landscape, dissolve altogether' (Ingold 2011b: 56). Here, storytelling is a mode of *dwelling* that brings the storied farmscape into being as the farmscape morphs from farmland into walks, conversations and embodied encounters. This chapter looks at a range of farmscape stories told by the farmers

© The Author(s), under exclusive license to Springer Nature Switzerland AG 2021

S. C. Haedicke, *Performing Farmscapes*, Performing Landscapes,
https://doi.org/10.1007/978-3-030-82434-1_2

themselves, spoken in their own words and embodied in their repetitions of agricultural tasks. The case studies highlight the varied ways of knowing these performing landscapes through the three modalities introduced in Chapter 1: *local farmscapes* by focusing on the significance of local, home-grown insider knowledge to understand the farmscape, *working farmscapes* by recognising the hard, and often invisible, work that goes into constructing farmscapes, and *naturalcultural farmscapes* by acknowledging the entangled collaboration of nature and culture in the formation of farmscapes. This chapter interrogates how the artists use these lived, place-based stories that braid the spatial and temporal not only to bring the audiences into the storied farmscape, but also to inform and engage the general public in issues affecting farming and the food system and take a first important step in establishing food citizenship.

Theatricalised Farmscapes: Theatre Passe Muraille's *The Farm Show* and James Rebanks and Chris Monks', *The Shepherd's Life*

Theatre Passe Muraille's*The Farm Show* (1972) and James Rebanks' autobiography, *The Shepherd's Life: A Tale of the Lake District*, adapted for performance by Chris Monks (2016), are very different theatrical pieces, but they both highlight working farmscapes that offer insider insights into local contexts. They each rely on the farmers' words, but simultaneously create a sense of community performance that grows out of extensive community participation. In August 1972, Theatre Passe Muraille premiered *The Farm Show* in Ray Bird's barn in Clinton, Ontario, Canada. Paul Thompson, artistic director of Theatre Passe Muraille, and a group of six actors left Toronto for a summer in the farmlands around Clinton. They lived and worked with the farmers and from their conversations and shared farm tasks, the actors devised *The Farm Show*. It was first seen by the farmers themselves, but it soon toured throughout Canada and internationally, often performed in auction barns rather than theatres. Theatre historian, Alan Filewod attributes the play's startling success to a pioneering creative process in which the actors collectively devised the script both from the words and farm tasks of the farmers and from their own experiences on the farms. He argues that this collective creation process inspired the development of documentary theatre in Canada

(Filewod 2008: 1–13).[1] In 1974, Michael Ondaatje filmed 'The Clinton Special: A Film about *The Farm Show*' by working with the original actors and participating community members to 'stage' the devising process and document scenes from current touring performances. And, in 1999, Michael Healey returned to *The Farm Show* as a stimulus for his award-winning play, *The Drawer Boy*, that draws primarily on the experiences of one of the actors, Miles Potter, and on Ondaatje's film. In the Introduction to the published version of *The Farm Show*, Paul Thompson wrote:

> The idea was to take a group of actors out to a farming community and build a play of what we could see and learn. There is no 'story' or 'plot' as such. The form of the play is more like a Canadian Sunday School or Christmas Concert where one person does a recitation, another sings a song, a third acts out a skit, etc. Nevertheless, we hope that you can see many stories woven into the themes of this play and that out of it will emerge a picture of a complex and living community. (Theatre Passe Muraille 1976: 7)

Living in a vacant farmhouse for six weeks, the actors spent every day meeting the farmers. One actress explained, 'We were supposed to go to meet people, and that was very hard 'cause we were shy and they were shy of us and it was a bit awkward' (Ondaatje 1974). Ondaatje's documentary reconstructed one of these initial 'awkward' meetings between the actor, Miles Potter, and the farmer, Mervyn Lobb:

> Miles: (climbs the ladder to the barn loft) Hi, my name is Miles Potter.
> Mervyn: (continues to move bales of hay) Oh, yeah.
> Miles: I'm from Toronto.
> Mervyn: Oh.
> Miles: I'm an actor.
> Mervyn: Oh, yeah.
> Miles: We're up here, you see, because we're going to do this play about farming.
> Mervyn: (laughs)
> Miles: And as part of it, we thought we'd go around to different farmers and see if we could be of some use.
> Mervyn: (grunts) Uh.

[1] See also Filewod (1987) and Usmiani (2008: 29–37).

Miles: (clears throat) Can I be of some use? (Ondaatje 1974)

To devise the play, the actors adapted the conversations, mannerisms and farm tasks to present 'a living portrait of a farm community' (Nunn 2008: 18). The published version playfully acknowledges the centrality of the farmers' contributions to the production in its dedication to them: 'All the characters in the play are non fictional. Any resemblance to living people is purely intentional' (Theatre Passe Muraille 1976: 15), and in the initial public performances in the farming community, the farmers recognised themselves in the characters. Les Jervis praised the show, saying 'this boy, Fox, he mimicked me pretty good' (Ondaatje 1974). The actors not only performed the people they met however; they also performed themselves getting to know and care about these people, and they made fun of their own naivety. As a consequence, the local context of *The Farm Show* was not just the farming community, but also the urban–rural encounters and the actors' awareness of how they were seen by the locals. Their performed presence as the people-who-went-to-Clinton never faded from the performance.

The Shepherd's Life began as James Rebanks' autobiographical account of his lived experiences and embodied sense of belonging on his sheep farm in the Lake District in England. Rebanks wrote his autobiography as a long letter to his dying father. The book, *The Shepherd's Life: A Tale of the Lake District* (first published in 2015), was an immediate success. Adapted by Chris Monks, Rebanks' life-stories were transformed into a community performance that implanted his lived experiences into the performing bodies of both professional actors and community cast members. The spoken text in the play was, for the most part, drawn directly from the autobiography's written conversations, and the character James Rebanks (played by professional actor Kieran Hill) acted as narrator of his own life story speaking more philosophical sections of the book directly to the audience with only occasional forays into the scene itself. Hill did not try to 'become' Rebanks but rather told his story. The sheep and sheepdogs were puppets often manipulated by the community participants. *The Shepherd's Life*, performed on the thrust stage at Theatre on the Lake in Keswick, Cumbria, UK, in spring 2016, relied on an interdisciplinary collaboration between artists and farmers to translate the local farmscape and its everyday sheep farming practices into a community performance that drew attention to the labour needed to maintain its iconic rural landscapes and to competing attitudes towards centuries-old farming traditions, the land and its uses, and its complex rural ecologies.[2]

Both *The Farm Show* and *The Shepherd's Life*, nearly two generations apart, place the actual words of the farmers into the mouths of actors, giving the plays a sense of autobiography as they use the farmers' stories to reveal insider knowledge of the local farmscape. And while neither of the pieces was site-specific, they both highlighted the significance of the local context through re-enacted farming tasks and visual imagery. When the actors took *The Farm Show* on tour, they recreated the strong sense of the local felt in the initial performances in the farming community around Clinton, Ontario, by covering the stage floor with a large map of the roads and farms around Clinton. An actor opened the show with 'So what you have is a sort of map of the community—the community we lived in and the names of the people we got to know' (Ondaatje 1974). Walking around the map, he pointed out the location of the farmhouse where they stayed for the summer and the farms they visited, and so established an intriguing link between local and located that acted paradoxically to draw the audience into the lived farmscape and, simultaneously, to distance them by positioning them as detached observers viewing points on a map. In the performances beyond Clinton, the actors also used borrowed farm implements: 'a few things from Clinton. This is part of a bean dryer. These are straw bales, not to be confused with the hay bales you'll hear about later. An old cream can. Some crates. An actual Clinton shopping cart' (Theatre Passe Muraille 1976: 19). In *The Shepherd's Life*, Monks also used visual cues to highlight the local landscape as he relied on an artistic interpretation of the actual farmscape eminently recognisable to the farmers. The set, for example, was dominated by a tall step-like structure of multiple levels echoing the Lake District's fells. Scenes taking place in other locations (a local pub, an agricultural show, Oxford University) were created by single pieces of furniture or fences set in front of the fells. Three large screens, mounted in a U-shape along the back wall, were used to project photographs or videos of the Lake District, often

[2] I saw this production in April 2016. On 3 June 2017, I returned to Keswick and interviewed Tess Ellison (Production Assistant) and several community participants: Ruth Hellen (puppeteer), Derek Cannon (member of the chorus), and Hannah Holmes (performed Rebanks' daughter Bea) and her mother. Cannon, discussed the challenge of 'bringing the outside into the theatre' and stressed the importance of enabling the audience to believe there are real sheep and sheepdogs on stage (Personal interview, 3 June 2017). Ellison also shared an in-house video of the performance with me.

to locate the theatricalised locale within the actual geographic location. In the scene about the impact of the UK Foot and Mouth disease on Rebanks' farm in 2001, for example, the family mourned the loss of the flock as they watched the billowing smoke as the sheep's carcasses were burned in distant pastures on the screens. The screens were also used to reveal the *art* of the jobs on the farm by juxtaposing a video of a farmer doing a particular task onsite with the same task enacted on stage by the actors, like building a stone wall, shearing a sheep, training a sheepdog, preparing sheep for an agricultural show or the gathering—the herding of the sheep from the peaks to the farms below.

In both *The Farm Show* and *The Shepherd's Life*, the modality of the local was interlaced with the modality of the working landscape as re-enacted farming tasks, sometimes realistically recreated, sometimes more choreographed, constituted much of the dramatic action. One of the most memorable scenes in *The Farm Show* was the 'Bale Scene' created by the actor Miles Potter (whose awkward first meeting with the farmer Lobb was already discussed). Potter went to Lobb's farm to help store the hay in the barn loft (the 'mow').[3] That farm job included loading the bales from the wagon onto the chute ('outside loader') that transported the bales up to the mow and then taking the bales off chute and stacking them in the mow. Potter humorously recreated his difficult experience as he performed his attempts to do this job and made fun of his own ineptitude as he mimed picking up and dropping bales, tripping over twine and falling in holes in the mow:

> I picked up the bale, and put it on the loader… Look out! (*apologetically*) Fell off… Well, I picked it up and put it on the loader, pressed it down, and up she went! … oh, yeah, … another one. Well it was right about this time that I figured out one thing. I wasn't dressed for it. You see, all I had on was a little pair of shorts, and the only way I could get the bale up onto the loader was by *pushing* it along with my leg, which was getting pretty cut up, and sore. Well, it was pretty hot that day … about eighty-five, and we were working at what I figured was a pretty good pace—look out! (*bales come crashing down*). (Theatre Passe Muraille 1976: 40. Ellipses to indicate pauses and italics in original text)

[3] This scene is performed in full in Ondaatje's *The Clinton Special* (1974).

He offered to unload the bales in the mow, thinking it had to be easier, only to find it was even harder. Running around the stage space, enacting the ways he lifted or rolled the bales, he finally thought the job was finished:

> And, finally. After what seemed like forever! There it was. The last bale. (*Bales have been getting heavier. This one is heaved with great effort.*) I went… and stood in the entrance to the mow. I let the breeze blow over my body…, and dry the sweat …, and the blood. (*Pause*) Then I looked *down* … and they were bringing up *another wagon*! (Theatre Passe Muraille 1976: 43)

He finished the scene with 'Now I ask you … why? Why would any human being *choose*, for the better part of his life, *twice* a year, to put himself through that total and utter hell? I didn't understand it then … and I don't understand it now' (Theatre Passe Muraille 1976: 43). In Ondaatje's film, Potter expressed his real anger at Thompson for conning him into that experience even though it became one of the most popular scenes in the play.

The innumerable examples of creatively re-enacted farm tasks not only performed farming practices, but also revealed a tension between realistic representation and an awareness of the need for theatrical fiction. That was most noticeable in the scenes where the actors became farm machinery; for example, in the scene entitled 'Man on Tractor' (I, ix), three actors created the tractor (the high seat and two large tyres) and made all the tractor noises. Potter sat on the shoulders of the central figure and drove the tractor as he gave a monologue on the relationship between the farmer and his tractor: 'Now the thing about a farmer and his tractor is that he's gotta spend so much time on it.… It's like my friend Bill Lobb used to say… sometimes a man gets to taking better care of his tractor than he does his wife!' (Theatre Passe Muraille 1976: 62). Sociologist Michael S. Carolan also found this bond between farmer and tractor as he researched Iowa farmers. One farmer he interviewed explained that he knew the land through the seat of his tractor: 'When you've been doing this as long as I have you can tell a lot about the ground by how the tractor handles. It's almost like the tractor is part of me' (Carolan 2008: 413).

The Shepherd's Life also relied on the re-enactment of farm work as its dramaturgical structure followed the annual seasonal rhythm of a working sheep farm with its various shepherding tasks associated with particular times of the year as in the original autobiography. Superimposed on this work cycle were events from Rebanks' life drawn from several decades, but associated with a particular season. These memories punctuated the seasonal rhythms and highlighted both their uniqueness in the life-cycle and their ordinariness in the seasonal cycle. For example, in the section 'Autumn', James explains the importance of the annual agricultural show: 'Everything that makes us who we are culminates in the autumn' (Rebanks 2016: 108). The adult actor playing James describes a memory of when, at nine, he successfully completed his first sale of a tup at an agricultural show to Jean, a 'born-and-bred sheep woman …[who] strikes a hard bargain' (Rebanks 2016: 110). This memory is then enacted by a child actor playing the young James.[4] In 'Spring', James describes the seasonal work through a specific memory of lambing as the character James encourages his second daughter Bea, only six at the time, to deliver a lamb just as her older sister had a couple of weeks earlier. Together, with the help of a sheep puppet, they perform the birth.[5]

Rebanks' autobiography sought to make visible this 'farming way of life [that] has roots deeper than 5000 years into the soil of this landscape' (Rebanks 2016: 2), often expressed in the production by character James as narrator. Sometimes didactic, these sections nevertheless highlighted the political import of the contrasting views of the landscape as productive farmland or iconic rural scenery. Early in the book and the play, Hill as Rebanks narrated a moment in school when he realised that the farmscape he knew and worked had another contrasting identity that had nothing to do with his working farm. This other landscape, clearly valued by his teacher, was 'a "wild" landscape, full of mountains, lakes, leisure and adventure…the playground for an itinerant band of climbers, poets, walkers and daydreamers' (Rebanks 2016: xv[6]). It was a 'landscape of the imagination' (Rebanks 2016: xvi) devoid of the people who work the land. Unlike his teacher, Rebanks saw the Lake District as a 'peopled landscape… a place crafted by largely forgotten working people. It is a unique,

[4] See Gardner (2016) for a photograph of this scene.

[5] Rebanks' daughter was performed by Hannah Holmes, a young community actor who was a good friend of Rebanks' daughters (Hannah Holmes, Personal Interview, 3 June 2017).

[6] The performance dialogue closely resembled the written text, so page numbers from the book are given.

man-made place, a landscape divided and defined by fields, walls, hedges, dykes, roads, becks, drains, barns, quarries, woods and lanes. I can see our field and a hundred jobs that I should be doing instead of idling up on the fell' (Rebanks 2016: 3–4). In an interview on BBC Radio 4, he argued that seeing the Lake District landscape simply as beautiful and romantic is superficial, 'like a view you put on a chocolate box' (Rebanks 2015: n.p.). Rebanks' concept of the farmscape, so evident in Monks' adaptation, was inextricably tied to jobs around the farm rather than vistas; here, the land was worked and lived-in. 'Landscapes like ours are the sum total and culmination of a million little unseen jobs' (Rebanks 2016: 44), James insisted in both the book and the production. The farmscape is the farm work, not the view:

> Every wall, every field, every hillside, every hedge, and every sheep mean something to us. They represent work done, and things owned, and lives lived to make those things what they are. ... To reduce this to just scenery is to deny its true meaning and its interest and depth. ... To be embedded in a place like this lets you see a depth of things there that a tourist or visitor would miss. (Rebanks n.d.)

James' grandfather often expressed his disdain for tourists who walked the public footpaths on the fells, oblivious to the work that created the landscape. It infuriated him when these tourists viewed the real work of repairing the farm's stone walls as picturesque. In the theatre production, the grandfather taught the young James how to fit the stones together to build a wall as the sheep puppets grazed in the pasture. When the dogs barked at the hikers who appeared on the fell, the grandfather muttered 'bugger off. Leave us alone' and mimicked their comments about the beautiful day with sarcasm, pointing out they would be off the fells at the first sign of bad weather. Moments later when William Wordsworth and Alfred Wainwright entered the scene with theatrical flair, he berated them for how they ruined the farmscape he knew and loved with romantic notions of *scenery*. Rebanks commented later in words that interweave the modalities of local, working and naturalcultural farmscapes:

> I wondered whether the people on that mountain saw the working side of that landscape, and whether it mattered. In my bones I felt it did matter. That seeing, understanding and respecting people in their own landscapes is crucial to their culture and ways of life being valued and sustained. What you don't see, you don't care about. (Rebanks 2016: 88)

With knowledge and respect, Rebanks suggests, the non-farmer can become a food citizen by embracing the responsibility of Haraway's 'response-ability… collective knowing and doing, an ecology of practices' (Haraway 2016: 34). Haraway draws on the writings of Isabelle Stengers to establish a parallel between this response-ability and a relay of string figures, like cat's cradle, where the passive set of hands holds out the entangled string for the active set of hands to remake in the taking—back and forth. And from this response-ability, a conversation of listening and responding and a practice of caring can grow: 'matters of fact, matters of concern and matters of care are knotted in string figures' (Haraway 2016: 41). Through this production of *The Shepherd's Life*, audiences could take the string figure offered by the farmers through the actors and begin to discover and vicariously experience the 'dwelling' farmscape of a sheep farm. By understanding the local, working naturalcultural farmscapes in *The Farm Show* and *The Shepherd's Life*, the audiences acquire new knowledge that, in turn, encourages a sense of responsibility/response-ability as they move towards food citizenship.

Dialogues Within Farming Communities: Mary Swander's *Farmscape* and *Map of My Kingdom*

In her plays based on interviews with Iowa farmers, Mary Swander (Poet Laureate of the State of Iowa, 2009–2019) explores the potentially negative impact that a farmer's bond to the land can have on a sense of identity and self-worth in *Farmscape* and on family dynamics in *Map of My Kingdom*. Years ago, Yi-Fi Tuan claimed that a farmer's bond was especially strong because of a 'physical intimacy [and] material dependence' (Tuan 1974: 97) on the land, and his observations are still relevant. He quoted one farmer's description of what the land means to him: 'It's a part of me; it's as much me as my own arms and legs.… The land, it's friend and an enemy; it's both. The land, it runs my time and my moods; if the crops go well, I feel fine and if there's trouble with growing, there's trouble with me' (Tuan 1974: 97). Swander's plays address the traumatic experience when a farmer loses the farm or faces difficult land transfers from one generation to the next. The plays bring these issues into the open and encourage often difficult conversations within farming communities.[7] Her first play, ***Farmscape: The Changing Rural Environment***, began as a class assignment in Creative Writing at Iowa State University, United States, in 2007. She and the students collaborated on a verbatim play based on interviews with local farmers about the rapidly changing rural Iowa farmscapes as large-scale monocropping organisations acquired small family-owned diverse crop farms and the priority of profit often replaced care for the land. One of the farmers interviewed, Martin, had lost his farm during the Farm Crisis in the 1980s. When they interviewed him over twenty years later, his distress at his loss was still so vivid that he could recite all his farm equipment sold at the farm auction. That auction became the through-line of the play and a key to performing the lived experience of the Iowa farmscape. Martin explained:

> [Y]ou can't put a value on land that's been in the family…but that's always been the problem with farmers. Farmers have always developed a weird type of bond with an inanimate object, you know. There's a bonding that takes place between the land and the person that is almost the same as between a husband who loves his wife. There is. There really is. There's a strong bond there. (Swander 2012: 31)

Many of the farmers interviewed feared that 'strong bond' was being lost in the changing rural landscape. The play's dramaturgical structure allowed for the many different ideas and concerns of the farmers interviewed to be presented side-by-side almost like a living photographic album rather than a conversation or a causal dramatic structure. In each scene, the characters speak on the same topic, but rarely to each other. Instead, each remains within his or her own location at home or at work where the interview was originally conducted. This presentational format, reinforced by its staging as a play reading, highlights the verbatim nature of the play and makes the unseen interviewer's questions seem audible. The play's dramaturgy and staging certainly do not create realistic scenes and yet, paradoxically, the *reality* of the farmers confiding their disappointments and concerns directly to the spectators has a profound and quite *real* presence.

After its first performance in 2008, the 'lights faded. The lights rose again. Applause. More applause. The students and I stayed on the stage.

[7] See also Nebbe and Boden (2014).

The real interviewees joined us. We had a dialogue with the audience. The first *Farmscape* "talk-back" was born' (Swander 2012: 59). Although the play reading was supposed to be a one-off, its success led the Leopold Center for Sustainable Agriculture at Iowa State University to give Swander a grant to tour the production to three locations in Iowa, and within a day, the venues were booked. More bookings followed, and the show continued to tour for several years, always with a post-show 'talk-back', actually more a conversation, as an integral part of the performance. Swander explains:

> The talk-backs went to the real heart of *Farmscape*. The rural environment is changing quickly all around us in Iowa, the U.S. and the world. *Farmscape* raises key questions about these changes and about the stewardship of the land. What are the lingering effects of the Farm Crisis? What has happened to the family farm? How do we raise our crops? How do we raise and slaughter our livestock? ... How are minorities and women involved in the scheme of things? Who is making it financially? How are we using our natural resources? (Swander 2012: 63)

The lively, often heated, talk-backs with the local audiences primarily of farming families confirmed the significance of the issues raised by the performance and how they played out in a local community. The play was performed in universities, town halls, community centres and other informal non-theatre spaces, so audiences felt more 'at home' and post-performance conversations grew organically from the performance.

As in *Farmscape*, the farmer's bond to the land is the focus of *Map of My Kingdom* (2015),[8] but here the question is 'who's going to get the farm and what are they going to do with it? What is going to happen to the piece of land you've dedicated your whole life to after you die?' (Personal Interview, 28 December 2016). Practical Farmers of Iowa, an organisation that seeks to strengthen farming communities through worker-led investigation and information-sharing (a kind of participatory action research often associated with food democracy, discussed in Chapter 1), commissioned Swander to write a play on farmland transfer. Executive Director, Teresa Opheim, wanted to provoke much-needed conversations in the farming community about planning for what will happen to the farm once the farmer dies, an inheritance struggle that

[8] See Swander (2015b).

often results in familial conflict and financial problems. A provocative play and post-performance conversation seemed appropriate to Opheim. The issue is difficult to discuss, Swander muses, 'because you have to face your own mortality' (McCollough 2014: n.p.). One farmer said to Swander, 'Mary, it's a lot easier to talk about sex with your offspring than it is farmland transfer' ('Drama of the Land' 2015).

Swander admits that when she was first asked to write a play about farmland transfer, she wondered if there was 'drama' in the issue. Theatre Passe Muraille had introduced the topic in the final scene of *The Farm Show* where the farmer, his wife and three adult children stood 'in a row framed by a large picture frame' (Theatre Passe Muraille 1976: 97). As the parents reminisced about the farm that had been in the family for generations, their oldest daughter stepped out of the frame explaining that she got a job in London, Ontario and doubted whether she would return to the farm. As the parents continued listing the improvements they had made to the farm, their second daughter and son also stepped out of the frame. The scene ended with the 'painting' of the farmer and his wife being auctioned and sold (Theatre Passe Muraille 1976: 99). Unlike *The Farm Show*, Swander chose to introduce several land transfer narratives. She understood that to tackle the issue of who gets the family farm in a play that would spark conversations required a different dramaturgical structure than her earlier plays that relied on side-by-side interview excerpts. Here, she moved away from strict verbatim as she combined, reworked and fictionalised interviews for dramatic impact. For *Map of My Kingdom*,[9] Swander created a narrator who was the sole actress in this one-woman show and a narrative arc that followed the narrator's developing awareness of the complexities and significance of land transfer issues. The inspiration for this character, Angela Martin, came when Swander met Mike Rosmann, a psychologist, former farmer and land mediator who worked with families experiencing land transfer conflicts, and Angela too became a psychologist and land mediator. Her stage world is made of five file boxes holding papers from the case files she had worked on as a land mediator and a few books. Throughout the play, she interweaves stories from her various semi-fictional clients (all based on actual interviews) with King Lear's misguided attempt to divide

[9] Swander shared version 2 of this unpublished manuscript with me. See also Swander (2015b).

his kingdom among his three daughters as a leitmotif to highlight the intense conflict land transfer can cause.

The play opens with Angela explaining how her grandmother sold the family farm without telling the rest of the family: 'the farm—that was steady, you know—always home, no matter where we were, always there. … I puzzled over that transaction for years' (Swander 2015a: 6). As Angela moves through her clients' stories, she uses her own growing comprehension of why farmland transfer is so difficult to guide audiences towards a similar recognition of its importance for the family. Admitting that she is beginning to understand her grandmother's decision to sell the family farm outside the family, Angela holds a copy of *King Lear* and proposes that Shakespeare must have understood as well since the story of land transfer is Lear's story too: 'any farmer reading *Lear* could see that the whole problem in the play was this idea of the land, how it's passed on…transferred from one generation to the other' (Swander 2015a: 10). But *Map of My Kingdom* does not just reiterate the need for a plan to divide the kingdom as Swander ties that narrative of land transfer to one of stewardship of the land, evident in a story in the play, based on an anecdote told to Swander by Iowa farmer, Tom Frantzen. As a young farmer from a Catholic family, Frantzen turned on the radio to listen to the sermon Pope John Paul gave when he visited Iowa. The Pope spoke about the need to be stewards of the land and how we all need to leave the soil in better condition than how we find it: 'the land is yours to preserve from generation to generation' (Roth et al. 2014; Swander 2015a: 25). While the details in the play's scene differ slightly from Frantzen's telling of the story, the core of preserving, even improving, the land for the next generation is the same. In the play, Gerry and Marilyn also respond strongly to the Pope's words. While Marilyn goes to listen to the Pope live, Gerry stays on the farm to paint the barn as he listens on the radio. He admits later that he 'about fell off the ladder when the Pope talked about the land' (Swander 2015a: 25). For both Marilyn and Gerry, in different locations, the message of the need to sustain the land for their children is an epiphany, and they decide to make sure they all agree on the plans to pass the farm on. Angela lauds their success as the effective land transfer happens at Gerry's death. *Map of My Kingdom* ends on the theme of recognising the importance of the naturalcultural farmscape for the survival of the planet:

> It's a tough conversation on how we end our own 'little while' with the land, let somebody else's 'little while' start. We come and go, the land is always here, and the people who love it and understand it. . .we only have that little sliver of time to make sure that the land keeps getting loved, keeps getting understood, one 'while' after the other. (Swander 2015a: 27. Ellipses in original)

The play has led to extended post-performance conversations and inspired longer-term initiatives in areas of farmers' mental health and ecological justice.

In 2018, Swander and Fred Kirschenmann of Iowa State University's Leopold Center for Sustainable Agriculture launched AgArts (AgArts 2021), an ongoing initiative that uses the arts and conversations they initiate to promote agricultural sustainability and healthy agri-food systems in part by bringing farmers and artists together. In *Farm-to-Fork Tales*,[10] for example, the farmers themselves tell their own stories in an evening of local oral history. In 2020, Swander started a series of podcasts called *AgArts in Horse and Buggy Land* (Swander 2021) that tell poignant and often whimsical farmscape stories to inform and engage non-farmers in the daily lives of farmers. She does not present the stories as documentaries even though they are developed from conversations with her Amish neighbours and other local farmers in rural Iowa.[11] Using humour, Swander narrates realities of farming in ways that encourage the podcast audiences to listen:

> The Sheep mow the grass, and then provide a freezer full of organic meat for the winter. ... The day we take them to the slaughter house is grim, for try as we may not to get attached, we are still mindful of the weight of our deed. To cheer up ourselves and the sheep, we make their last day as festive as we can giving them extra corn and a little hay before their trip to the butcher's.... [Since there are only two sheep, Swander loads them into the back seat of the car.] In they go on their haunches, mesmerised and playing statue the way they do when they're sheared. We strap seat belts around their middles, plop straw hats on their heads, and offer them one last cigarette before we speed down the road. (Swander 2020, Season 1, Episode 2: 'Taking Livestock to Market')

[10] See Swander (2018a, b). See also Nebbe and Stanton (2018).

[11] See Nebbe (2020).

In the same episode, Swander narrates how she put turkeys and ducks together in one coup even though her Amish neighbours warned her against mixing species. The ducks and turkeys became great friends, but the ducks found a hole in the fence and escaped to swim in the nearby pond, 'but wherever the ducks went the turkeys followed'. The turkeys soon followed, but 'ducks swim, turkeys do not'. Looking out her window, Swander saw the drowning turkeys and rushed to rescue them with a butterfly net. In the ongoing podcasts, Swander invites a non-farming public to experience imaginatively the joys and the challenges of rural farmscapes, and acquisition of this insider knowledge marks a first step towards food citizenship.

Storied Walks Around the Farm: Performative Walks with Ffion Jones' *The Only Places We Ever Knew*, Charlotte Hollins on Fordhall Organic Farm and Sylvia Grace Borda's *Farm Tableaux Finland*

Storied farm walks offer the general public the opportunity to enter a farmscape, live its stories in farmer's words and sounds of nature, encounter the topography of the land and experience a visceral response to the multi-sensory stimuli. In these storied walks, the farmland and the farmer are co-authors of the *performance* of the farmscape for the walkers, and the farm work and the land's specific local features fold into one another to produce a local-working-naturalcultural farmscape. Non-farmers can come into contact, even briefly, with an embodied knowledge of the agrarian countryside that has the potential to start them on the path to becoming a food citizen (Fig. 2.1). As Sociologist Michael S. Carolan claims, encounters like these can enable us to get to know 'the countryside in the flesh' (Carolan 2008: 412).

Ffion Jones' *Woollying the Boundaries*, a series of three practical pieces that 'performed' sheep farming, invited non-farmers into the 'backstage' of the Welsh tenanted hill farm, Cwmrhaiadr, where her father Glynne and brother Owen kept a flock of close to nine hundred sheep. These three pieces, *The Only Places We Ever Knew* (2010), *Ode to Perdurance/Awdl Amser* (2012) and *Dear Mick Jagger* (2013),[12] were created as part of her

[12] See Jones (2014) for detailed descriptions and analyses. *Ode to Perdurance/Awdl Amser* is discussed in Chapter 3.

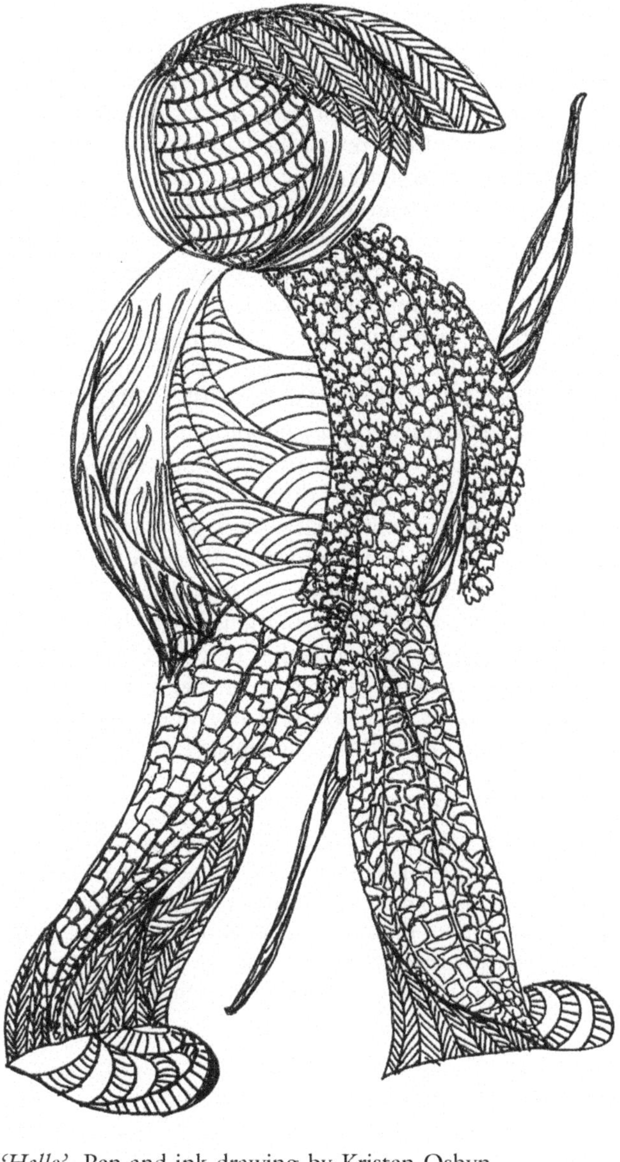

Fig. 2.1 *'Hello'*. Pen and ink drawing by Kristen Oshyn

PhD research, *Woollying the Boundaries: Perceptions of, and Interventions into, Upland Sheep Farming in Wales* (2014). In the first piece, *The Only Places We Ever Knew*, Jones created a storied walk around the farm that wove stories from past events and present-day realities into specific places in the farmscape. She sought to reveal the hard work on her family's sheep farm in Wales, labour often masked by the beauty of the landscape, and to counteract romanticised visions of farming that can get in the way of tackling pressing rural problems around the dangerous work, farm profits, land acquisition and tenancy, and sustainability. The storied walk clearly revolved around the two modalities of local and working farmscapes, but the naturalcultural farmscape modality was woven throughout with its underlying assumptions of 'becoming-with' the sheep and 'response-ability' to the land and livestock[13] (Haraway 2008, 2016). *The Only Places We Ever Knew* foregrounded place, people and labour to reassert the importance of 'lay narratives of the rural' (Jones 2014: 11). Her thesis opens with a memory that quickly dispels a notion of idyllic scenery as her brother yells to her to help wash a lamb he had found: 'The crows had him, and another one. I had to put the other one down' (Jones 2014: 9). Gently bathing the new-born lamb, she found that its anus and tongue had been ripped out. She held the dying lamb until her brother returned and 'then handed the lamb to him so that he could break its neck' (Jones 2014: 10). And in her unpublished text on her research process, *Woollying the Boundaries. Book 1: Fieldwork*[14] documenting the farm work in detail through photographs and descriptions as she shadowed her father, she narrates:

> Monday 29/3/2010. Weather—Heavy persistent rain… Morning: I arrive at the farm and Glynne talks about a set of twins that were born this morning: one of them is almost completely red (this is very unusual). He tells me that the mother of the twins was so frightened of her red lamb that she kept trying to run away from it. Glynne had to rub both of the twins together in order to make sure that they both smelled alike.… Afternoon:

[13] Jones uses these ideas as inspiration for *Dear Mick Jagger*.

[14] Jones created two unpublished books documenting her artistic process and practice: *Woollying the Boundaries. Book 1: Fieldwork* includes photographs and descriptions of the research process and *Woollying the Boundaries. Book 2: Artistic Practices* includes photographs and descriptions of the three practical pieces. Jones shared these books with me.

> Glynne wants to check on the red lamb. We find its mother and the white twin, but the red lamb is nowhere to be seen. (*Fieldwork* n.d.: 5)

They found the abandoned lamb, but it died a few days later. Dying and ugliness are as much a part of farming as birth and beauty.

Hundreds of photographs and vivid descriptions of daily life on the farm, like tagging the sheep, pulling lambs stuck during the birthing process, collecting the dead lambs from the pastures, bending the rams' horns if they are growing in towards the head, repairing fences, gates and farm buildings, or shearing the sheep, fill the extensive *Book 1: Fieldwork* to tell a narrative of farm life far from a 'simple rural life'. Using the farm where she grew up allowed Jones to rely on auto-ethnography as she drew on childhood memories and on participant-observation to gather her research. She explains:

> [M]y lifelong affiliation with the locale [her family's farm] leads me to suggest that at Cwmrhaiadr, places are many and variable: each of us has places that are special; each of us has places of differing importance. But, we also have collective places: places of memory, of knowing what we do there, and why we do it, places where our historical knowledge informs our attachment, places of hybridity, where place is made and maintained because of subtle interactions between human beings and natural phenomena. Daily activity re-captures, re-informs, re-makes, re-members and re-evaluates those places. Some are forgotten, only to be re-awakened by a new interaction. Places are subtle, unbounded and fleeting, but as a collective; they make up a strong attachment to the farm. (Jones 2014: 42)

These special places, she realised, are hard to share with outsiders since place is not just what one sees, it is what one does there, what one remembers and feels, who one is. How can those places and farm tasks be shared with an audience, many with little or no farming experience? How could the local insider knowledge about the farm be translated into the walkers' experiences? How could the farmer's intense bond with the land be performed? These were the questions Jones sought to answer with *The Only Places We Ever Knew*. Here, for a small audience of friends and colleagues,[15] Jones was able to foreground the specificities of the

[15] I was not in the audience, but Jones and I have exchanged numerous emails, and she shared the narrated text and her two unpublished books documenting her artistic process and practice.

farm work by linking anecdotes, history and memory to specific places on Cwmrhaiadr. She sought to enable the audience to experience daily and seasonal farming tasks, the repetitions of 'ordinary' activities rather than the big events. But I wonder if some of the ordinary farming tasks might seem to be 'big events' to the non-farmer who may never have seen a sheep being sheared or navigated the steep mountainous terrain.

Glynne acted as guide for the tour speaking through a megaphone to be heard over the tractor pulling the trailer of audience members up the mountain. He pointed out storied places along the way—a fifteen-acre field called '26 Acres', perhaps because of a previous tenant's exaggerations, and 'Cae Ffynnon', the 'well field', where bombs fell during the Second World War. At the top of the mountain near a small quarry, the riders became walkers. Placards with photographs of the many seasonal jobs done on the farm from *Book 1: Fieldwork* that Jones had taken during the first year of her field work as she shadowed her father marked their path down the mountain (Fig. 2.2).

Fig. 2.2 A placard on the mountain path. Ffion Jones, *The Only Places We Ever Knew*. 2010. Photographer: Heike Roms

At an early stop on the storied walk, Glynne pointed out 'Taid's stone' where his father's ashes were spread because:

> from here you can see where he was born, [and] where he was reared from the age of three.... When we put his ashes here, we buried a time capsule below his stone. It contains some information about him and some photographs. My mother also put an alarm clock in there. I'm not sure why, perhaps to wake him up or something. (*Woollying the Boundaries. Book 2: Artistic Practices* n.d.: 14) (Fig. 2.3)

As a quasi-theatrical event, the walking tour also 'performed' moments based on stories of the past. As Glynne narrated seeing a tree struck by lightning years before, her brother Owen surreptitiously set fire to a small tree. A second theatrical intervention accompanied the story of Taid who helped a neighbour, Lewis Lewis, move house and commented to his family later how strange it was that Lewis' face was almost black with dirt even though he lived by the river. As he spoke, the walkers could see a

Fig. 2.3 At Taid's stone in Cwmrhaiadr, the Jones' sheep farm in Wales. Ffion Jones, *The Only Places We Ever Knew*. 2010. Photographer: Heike Roms

figure on the other side of the river with his face covered in dirt. These two performance interventions acted as a testament to the 'home-grown' nature of the storied walk performed by the dwellers.

Jones sought to create a storied walk that would 'change or disrupt the audience's way of looking … would draw the audience's gaze away from the landscape picture, instead focusing on the places within it or the activities that had taken place there' (Jones 2014: 63). Through the place-based stories Glynne told, re-enacted 'scenes', photographs and walking the land, Jones hoped to create 'layers of experience, from the mundane to the extraordinary; the framing of everyday farm work through photographs which attempted to challenge the notion of a "landscape picture"' (Jones 2014: 65). The storied walk was in part inspired by LEAF's (Linking Environment and Farming) annual 'Open Farm Sunday' that enables members of the public to have access to participating private farms on one Sunday in June every year. The Food Ethics Council lauds this type of participatory, experiential approach to communication and knowledge acquisition as meaningful in developing food citizenship (Food Ethics Council 2017: 23). Jones wanted to 'open' her family's farm to non-farmers to replace bucolic views of farming with its harsher realities. But she also understood the very real dangers of the walking tour. She said of her family farm:

> [w]ith repeated exposure to this particular terrain we have 'feet for the place'—an embodied understanding of the nature of the ground underfoot: slippery horizontal layers of shale; mud; bog; deceptively deep and spongy spaghnum moss; woody, springy heather; undulating and pockmarked mountain-land; slippery pasture after heavy rain. (Jones 2014: 61)

As the walkers began the descent of the mountain called The Falls, Glynne warned them to take care as he told the story of one of the landowners who fell to his death (Fig. 2.4).

Navigating those dangers, however, helped embed the farmscape in the walkers' bodies. The storied walk enabled the audience to experience the dangerous terrain, throw stones down an old and very deep lead mine shaft, or learn about the family farm's history, and this embodied participation in the farm began to complicate a romanticised conception of farming with stories of hard work, birth and death. The tour ended in the 'Top Shed' where Owen sheared sheep and trimmed their feet and

Fig. 2.4 Walkers descending the narrow path down the side of The Falls in Cwmrhaiadr. Ffion Jones, *The Only Places We Ever Knew*. 2010. Photographer: Ffion Jones

Jones' mother served traditional cakes and tea. The storied walk gave the walkers insights into a working lived farmscape well beyond its stunning vistas, as it linked local specificities, shepherding tasks and care for the land and all its inhabitants.

In November 2014, Owen died in an accident and since then, the circumstances of the farm have changed. The farm was an inherited tenancy with strict succession rules that allowed the tenancy to be inherited only twice. Glynne inherited Cwmrhaiadr from his father, so one of his children could apply for a second tenancy at Glynne's retirement, but there were a number of 'tests' that needed to be met. Key among them was the livelihood test that required that the proposed tenant had been earning at least 50% of his or her income from the farm for the seven years prior to the current tenant's retirement. Ffion had not been working on the farm, so it was impossible for her to take on the tenancy herself. Consequently, Glynne had no option other than to hand in his notice. The estate owners have since sold the farm, and it is now being converted into a pheasant shoot. Glynne has sold most of the sheep, but he is preserving the genetic bloodline by keeping a small flock of one hundred sheep on land further down the valley. In a recent email to me, Ffion wrote that *Woollying the Boundaries* 'has taken on a new life as a record of my family "doing farming" and being together during a time when the family was complete' (Personal email, 11 September 2020).

Charlotte Hollins of Fordhall Organic Farm in Market Drayton, Shropshire, in England was one of the women in UK agriculture that Sarah Harper and I interviewed for our performance-as-research project, *Who's Driving the Tractor?* (discussed at the end of this chapter). 'I grew up here, my father grew up here, our family have been here for generations as tenant farmers' was almost the first thing she said to us. On the day we interviewed her (26 April, 2017), we were also able to participate in one of her guided farm walks that pointed out the beauty of the farm itself,[16] but simultaneously invited the walkers to experience the working farmscape in both its history and present-day accomplishments in sustainable farming practices and community engagement embedded in the land. Charlotte does not view her guided tour for local or international groups who are interested in further education, organic farming or community development as a theatrical event. And yet, this highly performative walk draws the audience into specific farmscape locations with vivid stories of its inhabitants and farming practices: both Arthur's pioneering 'foggage' system that enabled the farm to become organic many years earlier and its present-day community engagement activities. Charlotte performs its sophisticated dramaturgical structure that weaves the topography of the land that we physically experience by walking the farm with a sense of belonging to place that we vicariously experience both in the dramatic story of the fight to save Fordhall Farm from developers and in her anecdotes about places that hold pleasant and difficult memories[17]:

[16] In many case studies, I mention the beauty of the farms' landscapes. I will discuss the challenges of referencing this trope and the resulting tensions between appreciating these iconic rural landscapes and acknowledging the hidden labour that constructs them in more detail in the case study of Louise Anne Wilson's *The Gathering/Yr Helfa* in Chapter 3.

[17] See Hollins (2008) and the Fordhall Farm website, https://www.fordhallfarm.com/group-visitors/.

The stories, entwining local, working and naturalcultural modalities of the farm go well beyond mapping the farm or even mapping personal connections to place and, I propose, act as an invitation to the public as food citizens to participate in the working life of the farm. Charlotte's storied walk begins in the community room of Fordhall Farm:

> Welcome to Fordhall Farm. We are England's first community-owned farm, after we raised £800,000 in 2006 to secure the farm away from industrial development. We are also an organic farm and have been farming organically here since just after the Second World War, following the forward thinking of the late Arthur Hollins. Fordhall is now one hundred and forty acres of organic pastureland where we rear cattle, sheep and pigs for sale through the farm shop and butchery, but that was not always the case.[18]

She explains how the farm shifted from dairy to beef farming soon after the death of Arthur's first wife in the 1970s and that she and her brother were born 'at Fordhall in the 1980s when Dad re-married. He was 67 when I was born, and 69 when Ben was born – so that is what organic living does for you!' The family began to receive eviction notices from the landowner who planned to sell the land for development in the 1990s. As Charlotte and Ben were both still in school, and Arthur was getting towards his 80s, the farm started a ten-year decline. In 2003, the farm's disrepair gave the landlord the ammunition he needed to evict the family, and he served them with a final twelve-month eviction notice. As we look around the beautiful community room, it is hard to imagine its past shabbiness. Ben and Charlotte returned to Fordhall determined to see if there was anything they could do to save their family farm. They were able to secure a new short-term tenancy for eighteen months only twenty-four hours before the family were due to be evicted. In March 2004, Charlotte and Ben, only twenty-one and nineteen, respectively, took on the management of Fordhall Farm. Charlotte then tells us that she will pick up pieces of the story along the walk in response to particular sites on the farm. 'Off we go and remember no romping across the fields! The ground is uneven and there may be cow pats around!', she warns.

[18] All the quoted sections from the guided tour are from the tour the author took on 26 April 2017 or from the written text of the guided tour shared with the author by Charlotte Hollins in an email, 27 August 2019.

As the story resumes once we are outside, she shifts from the early history of the fight to save the farm to present-day farming practices and community engagement activities and the farm's varied locations themselves begin to contribute to the storytelling.[19] A key aspect of Fordhall Farm is its emphasis on establishing strong ties to its community owners and members of the non-farming public as its varied activities foreground public access to private land to help overturn the 'get-off-my-land' mentality that Charlotte expresses quite vehemently is too prevalent in agriculture. At the community garden, we learn about the care farm for adults with learning disabilities and projects for young people who struggle in school to help them develop self-esteem and confidence as they acquire new skills on the farm, like construction of the straw bale bunkhouse or the willow dome for outdoor community events. We pass the compost loo that, we were told, transforms human waste into compost-like material as long as the system effectively separates liquids from solids. Here, in her explanation of the symbiosis between the land and its inhabitants, Charlotte begins to introduce the third modality of collaboration of nature and culture on the farm, a recurrent theme throughout the walk. By the pig paddocks, Charlotte proudly introduces us to the approximately fifty Gloucestershire Old Spots, all free range and GM free,[20] but the only non-organic livestock on the farm. An old style bartering with Joules Brewery in Market Drayton circumscribes the pigs' lives: the pigs enjoy the non-organic waste hops that Fordhall swaps for pork. The Joules pub then turns the pork into sausage rolls given away to the customers at the bar on a Friday night. But it was not always that easy, Charlotte admits, as she paints a picture of the run-down farm in 2004 when they secured the eighteen-month tenancy. Charlotte and Ben realised that they needed to sell direct to the public in order to make the rent, so they took one of the six remaining pigs to the abattoir and brought back sausage. They put a chalk board at the end of the drive announcing sausages for sale and over the first weekend made £50. 'We thought we were rich!'. That first sale led to farmers' markets, hog roasts at private parties and an online shop.

[19] Some video footage of the storied walk, but no audio, is available in the *Who's Driving the Tractor?*, https://www.youtube.com/watch?v=Jc-jwkk6p_8.

[20] GM free means that the pigs do not consume any genetically modified products.

At a later stop on our walk on a high hill that affords a beautiful vista of the River Tern and the cattle pastures, a place that Charlotte calls the 'top of the valley', she explains how community ownership works at Fordhall Farm to guarantee its longevity. I think that the choice of this high promontory was not for its beauty however, but was a topographical confirmation that establishing community ownership through the Fordhall Community Land Initiative, a charitable society for community benefit, was a pinnacle achievement in Fordhall Farm's current success. This society succeeded in raising the £800,000 needed to purchase one hundred and twenty-eight acres of Fordhall Farm by selling over 14,000 non-profit-making community shares at £50 each to over 8000 people around the world and placed the farm into community ownership for perpetuity. So Fordhall Farm, that still sells community shares, is owned by over eight thousand people. Ben is the tenant farmer managing all the livestock, pastures, farm shop, bakery and catering trailers. He has a 100-year tenancy agreement with succession rights in it for his children if any of them would like to farm. Like any tenant farmer, he pays rent although not to a distant landlord, but rather to community members, many of whom participate in farm activities. Charlotte manages the other side of the farm business running the care farm, the youth projects, the café and special events, maintaining the farm trails, offering volunteering opportunities, leading school visits, running short courses, hosting weddings, renting out yurts for glamping and, of course, narrating the farmscape in the storied walks.[21] Charlotte and Ben's partnership in running Fordhall Farm not only makes it a sustainable agricultural venture, but also confirms that farm work goes beyond digging in the soil and caring for the livestock.

The next stop on the walk, the 'tree-planting area by the large field gate next to the stile' not surprisingly narrates the story of planting the idea for community ownership and opening the farm gate to the general public. Although this story preceded the community ownership story at the previous stop in chronological time, the farm's topography prevailed in the guided walk's dramaturgy. In 2004, Charlotte and Ben knew that they wanted to safeguard the farm for the long term and use it to help reconnect people to their food. By looking back at the farm's history, they could see that Fordhall was most successful when there were lots of

[21] In the case study of *Who's Driving the Tractor?* at the end of this chapter, I unpack the importance of community outreach in agriculture today.

people involved, and when this stopped, the farm lost its 'heart', as Charlotte says. They began to open their gate through volunteer weekends and anything else they could think of to get the local community onto the farm and worked with them to create a vision of a farm for the future. While these volunteers did help with tasks around the farm, Charlotte and Ben's goal was less about free labour than about breaking down the 'get off my land' syndrome that Charlotte so dislikes and that the community trust works hard to dispel. At a community meeting held at Fordhall Farm in February 2005, only a month after Arthur had passed away, they asked the local people what they wanted from this farm and by combining that with what Charlotte and Ben wanted from it as farmers, together they developed the community structure that exists at Fordhall today.[22] The approach regarded the non-farmer participants as citizen stakeholders investing in the farm because of what it represented and what its success would mean. Soon after that meeting, they were offered a first refusal to purchase the farm. After incorporating their new community structure, they had only six months to raise £800,000. Charlotte explains:

> In February 2006, two months into their fundraising we had raised an amazing £30,000 – more money than either of us had seen in our lives before. However, the old wiser volunteer board members were quick to point out, that whilst this was laudable, there were only four months left to raise the remaining £770,000! I admit that young naivety did us a lot of good here.

As more people volunteered on the farm, more people cared about its survival. One of the volunteers published an article on the 'Fight to Save Fordhall Farm' in *The Daily Telegraph* (Williams 2006: n.p.) with just over two months to raise £700,000. Several other newspapers picked the story up, including *The Guardian*, *The Observer*, *The Daily Mirror*, *The Daily Mail*, *Country Living Magazine* and even newspapers abroad. The response was incredible and in the final two weeks of the campaign, with £250,00 still to raise, sacks of post with cheques were delivered daily. Seventeen volunteers added information to the databases, recorded and banked all the cheques so that they were cleared by the deadline. Finally, only twenty-four hours before the deadline, the bank sent confirmation

[22] See Food Ethics Council (2017: 22–25) for a discussion this type of shift to a citizen mindset.

that enough funds had cleared, and on the 1st of July 2006, Fordhall Farm was placed into the community ownership of over 8000 community shareholders across the world for perpetuity. The story is inspiring and, as told by Charlotte in much more detail than I can include here, made us hold our collective breaths as the deadline rapidly approached. We felt the hard work, the sleepless nights, the anxiety and finally the joy in their achievement.

At a place on the farm called the Motte and Bailey site,[23] Charlotte explains her father's 'foggage' system (rotation of livestock, diversity in pasture and native breeds) that enabled the farm to become organic just after the Second World War. Again, there is a connection between a farm location and the narrative since a motte and bailey is a medieval system of defence and the foggage system of organic farming 'defended' the farm from the toxicity of chemical fertilisers and pesticides that had damaged its soil. The farm's landscape determines the dramaturgical structure of the storied walk as the episodes follow the lay of the land rather than chronology. The land holds memories and promise, and Charlotte becomes the medium through which the farmscape narrates its stories. This complex storied farm walk clearly demonstrates the significance of the local topography in understanding the farm, the ways in which the working farmscape contributes to its identity, and how the farm represents a naturalcultural environment of stewardship of the land, its multispecies inhabitants and the public.

One final example of an innovative storied walk where the topography of the farm plays a key role in establishing its narrative structure is photographer Sylvia Grace Borda's *Farm Tableaux Finland* (2015).[24] One aspect of this multifaceted project was the development of staged visual narratives of a range of farms, both sustainable and industrial, in the diversified agricultural landscape of Finland. Building on her earlier *Farm Tableaux* (2013) in British Columbia, Canada, she created the farms' pictorial narratives in Google Street View, in collaboration with John M. Lynch, Google Trusted Street View photographer, so the viewers

[23] Motte and Bailey refers to a medieval defence system where an area of land (bailey), encompassed by a moat or other barrier, surrounds a tower on a mound (motte).

[24] See Borda (2015a, b, c). See also Strom and Borda (2020).

could 'walk' around the scenes on the screen almost as though they were actually there. Google Street View gave Borda an opportunity to escape the limitations of a single static photographic image as she sought to dispel a romanticised view of farming as bucolic and instead to allow the farmers to tell their own stories. Creating 360°dioramas through Google Street View enabled the viewers not only to interact with what is seen, but also to experience a sense of intimacy with the places and people as they viewed farming activities in situ and entered agricultural settings not usually accessible to non-farmers. To create these unfamiliar virtual experiences, Borda invited the farmer to stage a real farm task for the camera lens. Each tableau included at least one farmworker in a position that reflected the farm work, a position held motionless for up to forty minutes so that the Google cameras could capture it. The animals, on the other hand, did not stay still but rather continued with their daily activities, so while they were certainly recognisable, they were motion-blurred—what Borda called 'ghosting'. Those images give the impression, I think, not only of a working farm, but of the interspecies' ***making-with*** of the farmscape. The ghosted reindeer on Hannu Lahtela's farm, for example, present a paradox of a still/moving photograph as their movement is visible in their blurred images. The farmers, not blurred, surround the reindeer for their round-up. This active moment of farm work visible in Google Street View is captured in a still image, much like Giacomo Balla's futurist painting 'Dynamism of a Dog on a Leash' (1912) that shows the blurred legs of a dachshund in motion. That still image, however, is animated as the viewer 'walks' around the farm, made possible by Google Street View technology. The farmland itself also becomes an active participant in these agricultural encounters in feeding the grazing reindeer and absorbing their blood when they are slaughtered for market. The tableaux vivants of this farm reveal the harsher realities of farming and initiate a process of knowledge acquisition for the viewers. Borda also sought to create a platform where the artist is witness to the evolving farmscape. To accomplish this, she placed herself off to the side in each tableau, camera in hand or sometimes just the shadow of the camera and tripod. But the viewers, too, become engaged witnesses to realities of varied farming experiences and co-producers of their farm experience as they 'walk' around the farm.

The tableaux, however, were only one aspect of the project as it also relied on conversations, collaborations and networks beyond the art to include farmers, food distributors, chefs, policy makers and others involved in agriculture in Finland:

> I intend to create a contemporary dialogue about farming to include farmers themselves, the consumers and other stakeholders … *Farm Tableaux Finland* acts as a reflector of social interpretation and ownership. Through this work, I provide a communication channel about the global challenge of achieving environmental and social sustainability. (Borda 2015b: n.p.)

By engaging the agricultural community in the photographic project and the audiences as witnesses, Borda hoped to provoke debate not only about farmscapes, but also about larger socio-political-economic issues inextricably linked to agriculture. And, in fact, farming networks (a key aspect of food citizenship[25]) developed around *Farm Tableaux Finland* and contributed to a draft of a Food Charter protecting consumers' right to access local foods and outlining national standards for food production to be amended to the Finnish constitution (Borda 2015a). The project uses art to inform the public about Finnish agriculture, but also to present a range of views about farming and, in turn, to provoke debate on sustainable agriculture in colloquia and exhibitions on her work and beyond.

These storied farm walks seek to enable non-farmers to experience agricultural landscapes through an embodied engagement, but it is important to think about how each one is mediated through its dramaturgical structure, its ways of telling the stories, and the content it shares. Each foregrounds how the working farmscape constructs the local specificities of the farm. Jones focuses the walk around Cwmrhaiadr on family experiences and foregrounds the intimate links to the land and sheep thus emplacing Ingold's observation that in a storied world, the relationships among the things of that world are the stories (Ingold 2011a: 160). Hollins relies on a trope of hard work and perseverance leading to success, and while she does not shy away from the struggles to save the farm, she frames them within that paradigm and so follows each challenge or disappointment with how it was overcome. And Borda's self-guided walks enable viewers to experience agricultural environments rarely open to the public at their own pace and from the comfort of their

[25] See Hassanein (2008) and Food Ethics Council (2017).

own devices, and they increase the potential reach of the project and its ability to augment the number of food citizens. Carolan asks, ‘how can we give non-farming bodies a feel for production agriculture’ (Carolan 2008: 419) and whether those encounters can influence their attitudes towards nature and farming. He argues that ‘it appears wise to think about ways to fold embodiments tied to agriculture into the world of non-farmers, particularly if we are serious about reducing rural tensions’ (Carolan 2008: 419). The storied walks where the farmscape can be felt by those walking and listening use the modalities of the local, working and naturalcultural farmscapes as a way to inform and engage participants and so can play an important role in shifting a food consumer into a food citizen who participates in the life of the farm and buys into its purpose. Food democracy comes into existence through informed consumers who are able to engage in dialogue on food issues and participate in collaborations with farmers, scientists and other pro-active consumers to advance sustainability. The storied farm walks move in that direction.

A Performance-as-Research Project on Gendered Farmscapes: *Who's Driving the Tractor? Conversations with Women in UK Agriculture*

Who's Driving the Tractor? was an exploratory, performance-as-research project[26] looking at the challenges and achievements of women in UK agriculture that I co-created with Sarah Harper (Artistic Director of Friches Théâtre Urbain) in 2017–2018.[27] The initial idea for this project began to take shape after I visited Five Acre Farm to see another project, *A PerFarmance Project West Midlands* (briefly discussed in Chapter 1). Five Acre Farm is a CSA (community supported agriculture) near Coventry, England, where the farmer is a woman. Becca Stevenson told me that she laughs aloud when new CSA members show surprise that

[26] The methodology of performance-as-research (PAR) or practice-as-research (PaR) has generated a significant amount of scholarly writing trying to define, or refusing to define, it. Notable examples include, but are not limited to, Allegue et al. (2009), Arlander et al. (2018), Kershaw and Nicholson (2011), and Nelson (2013).

[27] *Who's Driving the Tractor: Conversations with Women in UK Agriculture*, https://www.youtube.com/watch?v=Jc-jwkk6p_8. I have already discussed the interview with Emma Hamer (Chapter 1) and Charlotte Hollins' storied walk on Fordhall Farm earlier in this chapter.

she is the one who plants and harvests the vegetables grown on the farm. Curious about the implications of gender stereotyping in her anecdote, I tried a small, informal experiment. I asked a few friends, colleagues and students to sketch a farmer, and nearly everyone drew a male farmer. Yet articles I was reading about agriculture indicated that around the world, the number of women farming today is rapidly increasing. A World Bank study (2016), for example, claimed that globally, agriculture is 'feminising', meaning primarily that more women are farming. My initial research found that, indeed, the number of women farmers (the decision-makers on the farm in charge of the day-to-day operations) in both the United States and the United Kingdom has been steadily increasing over the past few decades so that now in the United States about 30% of farmers are women and in the UK close to 20%.[28] That percentage is significantly higher in Africa, South Asia and Latin America. In much of the world however, that feminisation of agriculture does not necessarily mean increased wealth, social mobility or access to land or credit for women.[29] I also found that many women work in agricultural roles beyond the farm as scientists, researchers and policy makers. In 2018, Minette Batters became the first female President of the UK's National Farmers Union (NFU) and was subsequently re-elected for a second two-year term in 2020. Several of the women Harper and I interviewed for the project also have important roles in major agricultural organisations as well as working on their farms. Ali Capper is now NFU Horticulture and Potatoes Board Chair, and Emma Hamer is Senior Plant Health Advisor for the NFU. Caroline Drummond is founder and Chief Executive of the UK's LEAF (Linking Environment and Farming), an agricultural charity that encourages sustainable farming practices and develops educational events and programmes, like the annual Open Farm Sundays when over 1600 UK farms offer farming activities and discussions to the general public (Personal interview, 9 May 2017). Susannah Bolton, Head of Research Knowledge Exchange at the UK's Agriculture and Horticulture Development Board (AHDB), develops knowledge-exchange programmes across all six of the AHDB's sectors that pass knowledge from agricultural scientific research onto farms that, she explained, gets the science into practice by 'identifying innovation and ensuring its uptake and the development

[28] See O'Neill et al. (2020).

[29] See Slavchevska et al. (2019).

of best practice on the farm and in the supply chain' (Personal interview, 28 March 2017).

I wanted to know how the women in UK agriculture identified themselves in relation to farming, if what they did on a daily basis differed from the men with whom they worked, whether they had faced barriers or not, what they would consider their triumphs, and whether (or how) they are changing UK agriculture. I became quite interested to see if I could develop what I learned into a performance that both looked at the changing roles of women in UK agriculture and unpacked the idea of a feminised farmscape. Around this time, I discovered Audra Mulkern's *The Female Farmer Project* and Marji Guyler-Alaniz's *FarmHer*. Both are ongoing multi-platform documentary projects of photographic stories and videos that give women farmers of all ages, mostly in the United States, a virtual stage to 'perform' their lived experiences and embodied practices on their farms. Mulkern argues that the story of agriculture has only been half-told as women are missing from the agricultural narrative although not from agriculture itself. Throughout history, women's names have often been erased from documents like land titles, bank loans or equipment purchases, and that erasure makes the quasi 'official' agricultural narrative a male-dominated document. Mulkern's project seeks to revise that misleading narrative. In the farmer portraits, photo stories and videos, the women write autobiographical sketches of their farmscapes: a rooftop farm in New York City, livestock farms from Seattle to Vermont, ranches in Arizona and more. Guyler-Alaniz's *FarmHer* seeks to create an online social community for women in agriculture and to inspire young women to choose farming as a career not only through the autobiographical narratives of the women she interviews, but also through a weekly television show about farming women and a series of podcasts going into more depth with the women interviewed in the project.[30]

These projects inspired Harper (co-creator and actress in the performance) and me (writer/dramaturg) to create a live performance-as-research project structured on principles similar to the online projects so that the women we met could tell their own stories in their own voices, but we could not expect busy farmers to rehearse and perform even for the three funded public performances that we were committed to. From the beginning, we knew that we did not want to create a one-woman

[30] These projects are current at the time of this writing.

show with Harper playing all the women we interviewed in a verbatim theatre piece or a more fictionalised play such as Graham Harvey's *No Finer Life* that we saw at the Pleasance Theatre in London in 2017. That one-woman performance told the real-life story of Elizabeth who joined the Women's Land Army during World War II and went to work on Frank and George Henderson's[31] farm in the Cotswolds. Playing Elizabeth and all the people she met, the actress narrated the challenges Elizabeth faced to prove her worth as a farmer. The play provided biographical information, but, as an audience member, I felt distanced from the real Elizabeth as she was mediated through the voice and movements of the actress. It was important to us that, in *Who's Driving the Tractor*?, the audience 'meet' the actual women farmers, hear their own voices, experience their own gestures and see their farms. Our goal was to have the women 'on stage' with Harper so that they could reveal, almost viscerally, their storied farmscapes themselves. To accomplish this, we videotaped all the interviews and spent time photographing the farm, the offices of AHDB and LEAF (Linking Environment and Farming), or the labs at the Warwick Crop Centre where the interviewees worked. The performance used the original interviews, carefully edited, so that the interviewed women on the screen seemed to be having a conversation with Harper, the live actress on stage. That embodied recreation of the original conversation in the performance enabled Harper to relive her initial surprise, concern or laughter at what she learned from the women.

Around this time, I also discovered *The Rise of Women Farmers and Sustainable Agriculture* (Sachs et al. 2016).[32] This publication looks at women farmers in the eastern United States. The authors asked the many women they interviewed how they got into farming, why they farmed, what the impact might be of their claiming identity as *farmer*, and whether their entrance into farming in increasing numbers introduced an alternative to conventional agriculture. Sachs had written a book in the 1980s entitled *The Invisible Farmers*, and she and her co-writers were surprised to find that, even in the twenty-first century, women farmers are still often invisible. What they found contradicted earlier

[31] George Henderson wrote the popular book, *The Farming Ladder*, in 1944. He and Elizabeth married, and she ran the farm after his death.

[32] See also Devine (2013), Little (2002), Sachs (1983, 1996), and Whatmore (1991).

analyses that claimed that women farmers were perpetuating an agricultural system that conformed to patriarchal norms. The women with whom they spoke were establishing innovative alternatives to existing commodity-based agri-food systems and marketing strategies in large part in response to gender-specific barriers facing women farmers trying to establish themselves in traditional agriculture in the United States. These barriers included limited access to land and capital as women rarely acquired farmland through inheritance and start-up costs for large-scale commodity food production are very high, agricultural machinery that does not fit women's bodies and requires significant upper body strength, and persistent sexism in agricultural institutions (Sachs et al. 2016: 6–15). These barriers often pushed women onto small, diversified farms and into non-conventional farming approaches, like organic food production. A consequence was that women farmers were often associated with environmentally and socially sustainable farming practices. Drawing on theories from feminist political ecology, the authors 'rejected biological essentialist notions of women's connection to nature, emphasizing that women's concern for the environment is connected to their daily activities, labor, and access to resources' (Sachs et al. 2016: 24). They argue that the women they interviewed were reshaping the agri-food system and going beyond agrarian feminism as defined by Jenny Devine. Devine claims that agrarian feminism among farm women in Iowa in the late twentieth and early twenty-first centuries foregrounds 'working partnerships between husbands and wives, women's work in agricultural production, and women's unique ways of understanding large-scale farming' (Devine 2013: 3). At the same time, this version of agrarian feminism relies on a 'politics of dependence' where the women recognise their dependence on their husbands, but use that acknowledgement to acquire legitimacy and access to public spaces like agricultural organisations, legislatures and agribusinesses where they can then speak with insider knowledge and authority. The women also draw on skills honed in coalitions of women in farming, like the Porkettes (affiliated with the National Pork Producers), since the 1970s. Devine argues that these farming women rarely identify as feminist, but use feminist models of conversation and collaboration. She calls these women 'stewards of the land, of resources, of food, and ultimately of democracy' (Devine 2013: 14). Sachs and her co-authors found a different form of agrarian feminism among women farmers in north-eastern United States that exhibited an increased pro-active agency.

To define this alternative agrarian feminism, the authors develop a 'feminist agrifood system theory (FAST)' (Sachs et al. 2016: 140–148). FAST focuses on thematic areas of gender equality on farms: claiming the identity of 'farmer' for women—a title that gives them legal access resources, land, equipment, and capital; developing alternative agri-food systems; confronting sexism in agricultural institutions; and establishing networks for women farmers.

I was interested to see if there was a similar pattern of changing roles, titles and activities of women in UK agriculture and whether we could tease out a *leitmotif* of agrarian feminism specific to the women in agriculture in the south of England. What Harper and I found was that the small sampling of women we interviewed identified as farmers or agricultural experts and dismissed the idea that there were significant barriers in their being women to success in their agricultural careers. As Susannah Bolton at AHDB said, 'the key is knowledge and experience, not gender' (Bolton, Personal interview, 28 March 2017). Each of the life stories of the twelve interviewees, we realised, was built on their shared assumption that knowledge and engagement in farming by the non-farming public, by food citizens, would benefit the agricultural sector in multiple ways beyond their significant purchasing power.[33] Food citizens can, for example, work with the agricultural community to secure reasonably priced food products both ethically sourced *and* economically sustainable for the farmer instead of low cost food produced at great cost to the health of farmworkers and the soil, and they can help seek government and corporate support for environmentally sustainable farming so that it is more profitable than unsustainable practices. A more informed and pro-active public can demand oversight and participation in the agri-food system to ensure that it is ethical and oriented towards the common good rather than corporate profit. While the interviewees explained their focus on communication and community engagement differently, we found that they all depended on a model of conversation and collaboration to ensure economic, environmental and social sustainability on their farms.[34] While some claimed that women were better at communication than men, most simply narrated their accomplishments that, in turn, highlighted

[33] The efficacy of purchasing power on affecting changes in food chains is explored in Chapter 6.

[34] See Jones (2020) who discusses women farmers' three C's: communicate, connect, collaborate.

what they had done to close an information gap between farmers and non-farmers through community engagement, knowledge exchange and conversations. Charlotte Hollins' community outreach work on Fordhall Farm highlighted the mutual benefits arising from opening the farm to the community (discussed earlier in the chapter), and Susannah Bolton explained the AHDB programme of 'monitor farms' to promote knowledge exchange between farmers and researchers. Becca Stevenson highlighted the role of CSAs in connecting the public to where their food is grown as did Caroline Drummond's work on Open Farm Sunday and other pedagogical events that LEAF organises. Ali Capper framed collaboration in terms of marketing strategies for British beers, and Rosemary Collier reminded us of the importance of teaching. We also learned that women in UK agriculture had well-established networks, like Women in Farming Network, Ladies in Agriculture, Women in Dairy, Ladies in Beef and Ladies in Pigs, that often sponsored events for the public. Among the women we spoke to, we found that the active promoting of this crucial, albeit often overlooked, communication between the agricultural community and the general public was an important aspect of what agrarian feminism means in the UK. In addition, it represents a significant step towards establishing food citizenship.

Who's Driving the Tractor? was performed in three very different locations. The first performance took place in the common room of Fordhall Organic Farm in Shropshire for the local community consisting of neighbouring farming families, local journalists and volunteers and friends of this community-owned farm. The second performance took place at a conference organised by University of Warwick's Food Global Research Priority,[35] one of the interdisciplinary research communities at the university. The audience here consisted mostly of academics in Life Sciences, Social Sciences, and Theatre and Performance Studies. The final performance took place at the National Farmers Union (NFU) Headquarters in Stoneleigh, Warwickshire. Here, the audience was made up of mostly NFU employees.

The performance of *Who's Driving the Tractor?* relied on four different modalities to reveal what the women we interviewed thought about women in UK agriculture. One modality was, of course, the autobiographical storytelling by the women themselves. This was accomplished

[35] For information about the Food GRP, see https://warwick.ac.uk/research/priorities/foodsecurity/.

through careful editing of the interviews so that the original interview questions could be asked by Harper live on stage and answered by the women who appeared on one of the two large screens behind her. Since the women were on their farms or in their labs or offices in the original interview, the audience 'visited' them in their own settings. This strategy offered the audience a sense of the original interview through the voices and gestures of the women, and the precise editing of their actual words created what seemed to be a 'live' conversation with Harper who admitted that she often felt that she was back in the room with them. The second modality was a *visual* storytelling that offered a theatricalisation of the conversation between Harper and the women, sometimes giving additional information, sometimes playing with the women's storytelling. This modality was achieved through projected photographs or videos of tours they gave us of their farms (inspired by the storied walks discussed earlier in this chapter), hand-drawn sketches, Harper's art-based activities creating a sculptural story in real time that was projected onto one of the screens as well as activities 'performed' by Harper. Most of the time, the images on the two screens were different, so, for example, the women being interviewed appeared on one screen with photographs of her farm, historic photographs or paintings of women in agriculture through the centuries on the other. Another combination was the interviewee or videos of her farm on one screen while a sculptural story unfolded on the other screen. I will give examples of this modality later in the chapter. Together, the stories in these two modalities provided the women we interviewed with the opportunity to present their own versions of agrarian feminism that foregrounded their farmscapes, agricultural skills and strategies to engage non-farmers through a range of socially-engaged and knowledge-exchange activities. The third modality was the performance of our learning curve through the project—part real, part fictional—that mocked our urban naivety, played with our penchant to imagine farming as a bucolic escape, gave farmers in the audience the chance to laugh at/with us and encouraged the non-farmers in the audience to join us in our acquisition of knowledge about women in agriculture. That modality was inspired, in part, by *The Farm Show* (discussed earlier in the chapter) that, as Filewod noted, 'is not simply about an Ontario farming community. Rather, it is a play about the experiences the actors passed through in the course of researching the material. The performance documents the actors' growing consciousness as they make sense of the lives of the farmers' (Filewod 1987: 35–36). The final modality was the performance itself where the interviewees and other members of the wider agricultural community could participate in a communal gathering to hear their own stories that they knew better than the artists, but that we told back to them through often unexpected means. A post-show discussion, inspired by Swander's model (discussed earlier in the chapter), was an integral part of the performance itself as one of the original interviewees whom the audience had just gotten to know on the screen, stepped forward to lead the live conversation. These modalities were interlaced to create a multidimensional gendered farmscape. Woven into this web of modalities in the project were leitmotifs evident in the other case studies: the significance of foregrounding the local and the importance of recognising the farmscape as a working landscape that depends on a collaboration of nature and culture.

The modality of our journey as artists provided a dramaturgical pathway through the varied farmscape stories. In the script, we created a somewhat clunky theatrical device for our learning process: Harper, after welcoming the spectators often individually as they took their seats, introduced herself as being in theatre for years (true), but always wanting to be a farmer (only true in a romanticised version of her life):

> Hello and thank you for coming. I've loved farms ever since I was a child. I used to have a toy farm. Well, everybody did, didn't they? I've been in theatre for ... oh ... decades, but I've really always wanted to be a farmer. I just love the countryside. The fresh air, the sheep grazing on the fells, the scent of newly mown hay, the hooting owl, the frost crunching under your feet on a February morning. It makes me... well it makes *me* want to read poetry. And as it happens, I've got one here. If you don't mind, I'm going read you a bit.

And she launched into a verse of Elizabeth Barrett Browning's poem 'The Herefordshire Landscape'. Her teasing of our nostalgia for a quintessential countryside caused many laughs among the farmers in the audience that increased as she asks them if she could be a farmer since she listens to *The Archers* and watches *Farming Today* and *This Farming Life*. As Harper continued to play with our naivety, she presented agricultural statistics in a 'professional' PowerPoint, only the charts and diagrams were cartoon-like drawings that introduced our modality of visual storytelling. The statistics made her question whether farming was a bucolic return to nature or really just hard work. But the statistics also showed that, in spite of the challenges, women were becoming farmers in increasing numbers, and so, in a double-take, she asked, 'But, wait a minute— what's going on here? Are the women nuts? If it's so difficult, why are more and more women entering farming all of a sudden?' It is at this point that the interviewees intervened as they appeared on one screen to take apart her fantasy and highlight the 'invisibility' of farming women that Sachs and her co-authors discussed (Sachs 1983; Sachs et al. 2016). 'Women have always been involved in farming, particularly on family farms they have been central in terms of whether that's feeding the men for harvest or whether that's helping out milking right through to driving tractors' insisted Caroline Drummond. Sarah Pettitt agreed: 'There's always been lots of women involved in farming. You just haven't seen them. You know, the housewife that does the books and the accounts, the daughter that does the mucking out. There's women in the background. Frankly, with all due respect to the guys that work so hard, so, so hard, they are supported by women'. The modality of visual storytelling placed these assertions in a larger socio-historical context as paintings through the ages of women planting and harvesting in the fields and orchards and tending the livestock were projected on the other screen. Freida Pettitt, Sarah Pettitt's grandmother, joined in reminiscing that when she started farming during World War II, it was mostly women because the Land Army women farmed to feed the nation in wartime. As she described the Land Army, historical photographs of these women at work replaced the paintings.

But, in spite of the numbers of women in agriculture, farming has long been identified as men's work, Harper mused. Men are the farmers, women are the farmer's wife or daughter. She asked the women whether there were gendered barriers they had to overcome in such a male-dominated profession. The women replied in a chorus of no's. Becca Stevenson, farmer at Five Acre Farm, ridiculed the notion of barriers by reducing it to clothing manufacturers who do not make farm clothing for women except in flowery fabrics with small pockets so they have to buy men's clothing much too big so that they can 'fit their bums in' (Stevenson, Personal interview, 10 May 2017). Marion Regan, of Hugh Lowe Farms, a large strawberry farm in Kent, laughed at a man's incredulity as she told the story of one of her female tractor drivers who was mowing a hilly field and noticed someone taking pictures of her. Concerned that she had not set the tractor properly, she returned to the barn to check her settings and saw the man taking pictures. She asked him what she was doing wrong, and he answered, 'Nothing! I've just never seen a lady tractor driver before!' (Regan, Personal interview, 25 May 2017).

As the women narrated more detailed autobiographical farmscape stories, the visual storytelling became sculptural, and our learning process receded into the background. The women took over the narrative. Harper asked Emma Hamer how she got into farming and as she explained that she did not come from a farming family, but when she was five, she received a toy farm for Christmas on one screen, Harper used the live feed camera to project her building of a toy farm with small wooden animals, farm buildings and fences on the other. As Rosemary Collier, Alice Midmer and Freida Pettitt explained how they entered farming, the toy farm grew (Fig. 2.5).

Harper also 'performed' farming activities that the women described. As part of this sculptural storytelling, early in the piece, she asked Sarah Pettitt what she wears when she works in her fields. As Pettitt, on one

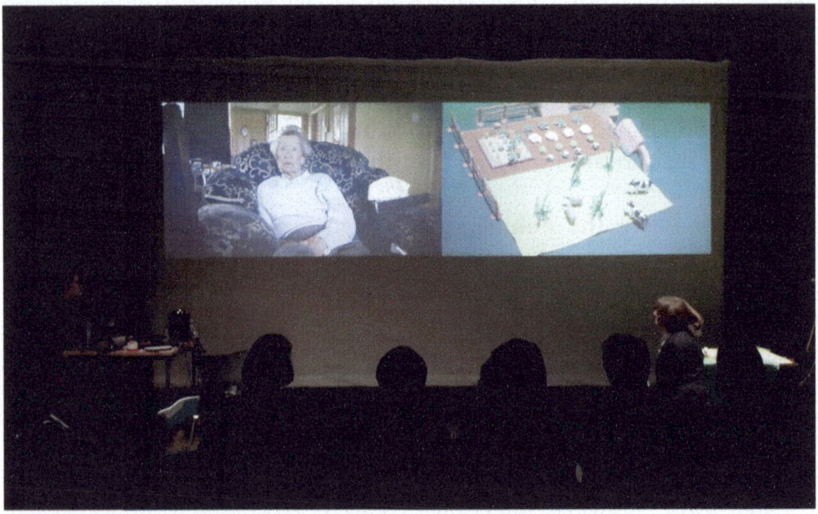

Fig. 2.5 *Who's Driving the Tractor? Conversations with Women in UK Agriculture*. 2018. Performance photograph showing sculptural storytelling on one screen; Freida Pettitt on the other. Performer: Sarah Harper. Photographer: Susan Haedicke

screen, mimed putting on her coat, scarf, hat and boots, Harper actually put the clothing on, but when she pulled on warm gloves, Pettitt chided her—no gloves. Harper stopped and turned directly to the screen to question why. Pettitt answered that there is no point since you have to take them off and put them on, off–on, off–on. Ali Capper told us that a freeze of her hops crop one year forced a decision about whether to abandon the crop since it never made much of a profit. She convinced her husband that they should continue to grow hops and that she would use her marketing expertise to build a stronger consumer base for British hops. As Capper told her success story in changing how people think about British beer by rebranding the product, Harper took on the role of her promoter as she displayed the British hops logo Capper developed and other items discussed in her story through the live feed camera. She then set up a beer tasting of sips of beer made at the brewery on Capper's farm that were passed around to audience members to taste. Or as Marion Regan explained the process of examining the strawberries for market, Harper whimsically recreated that process of measuring size and checking colour, scent and taste with hand-drawn charts and simple rulers. The visual storytelling, I think, helped the audience hear the many stories of the achievements of these women farmers, achievements that always involved some sort of communication and knowledge exchange. Like an illustrated novel, the images of Harper's enacted activities, performed or projected onto the screen, worked with the spoken text to provide a multidimensional view of agrarian feminism. While they added an element of whimsy to the often detailed narrative, they also emphasised the enormity of the accomplishments.

One final example of visual storytelling was the scene where Sarah Pettitt spoke with pride about her contributions to agriculture even though her father told her 'farming's not for girls'. When she was quite young, she introduced a new crop not only into the family business, but also into the British diet: purple-sprouting broccoli. She told the story in her own animated words[36]:

> I got some seeds and I, with the help of father and others at the end of the planting run, set a quarter of an acre of purple-sprouting broccoli. I

[36] Sarah Pettitt (formerly Sarah Dawson) tells this story in *Who's Driving the Tractor?*. For a video of the performance for the Food GRP, see https://www.youtube.com/watch?v=Jc-jwkk6p_8.

> nurtured it like a baby because nobody else was interested in it; … they thought it was just a waste of time. I had to find somebody who was interested in buying it and, through a local distributor, sold it to some of the top restaurants in London, such as The Ivy. I harvested it myself, packed it myself, and off it went. And it was successful, so the following year I grew a little bit more and the following year, I grew a little bit more. I planted it, and I nursed it, I harvested every single stalk, packed it. At this point, I had to get father and brother involved because it was rather a lot and I needed help in packing and wrapping and doing different things. This went on for about three, possibly four years. And there was interest when I was talking, you know, as I do—I'm good at that—I do a lot of that. And then one of the supermarkets said 'well, we might be interested in that.' I said okay and so we grew a little bit more, and I supplied the supermarket. It went really well so we grew a lot more, again still to the annoyance of my father particularly because he couldn't see the point. 'The mainstay of the business is cauliflower, broccoli, cabbage' he said. 'You know this is what we do. We grow hundreds of acres of this stuff. You want to mess about, causing me aggravation with this poxy queer gear that takes a lot of looking after. You deal with it; you just deal with it'. But now it was getting to a point where I couldn't just deal with this on my own, so I got my father and brother in there harvesting … When I was harvesting the crop, I got the highest yields that we've ever had since. The supermarket really liked it. I put it in other supermarkets, so I grew the business, grew the business, grew the business, and with my father and brother's support, grew the business, at which point I had to literally hand it over to them because we're talking fifty, sixty, seventy acres. It became a relatively important part of the business though we were still producing sprouts, cabbage and broccoli. It was perceived as a new vegetable—it never is, it's an old vegetable—but it's a new vegetable in terms of the exposure it got to the general public through the supermarket. It really took hold and people really understood the intrinsic health values of it as well and it's delicious, and so it grew exponentially. Purple-sprouting broccoli was in all the supermarkets, and today, it's the mainstay of the business. We stopped growing cauliflower, we stopped growing broccoli, we stopped growing cabbage, we stopped growing sprouts, and now we wholely, solely focus the small family business on purple-sprouting broccoli.

As Pettitt ended her story, Harper asked, 'Is your father proud of you?' After a brief pause, she slowly responded 'In many ways, yes. Ummm, of course'. She paused again, looking pensive and then added with a small smile 'But then, it's not something you hear very often'. In the performance, as Pettitt told this story of her success in growing and selling purple-sprouting broccoli with vivacity, humour and elaborate hand gestures on one screen, the farmscape came to life on the other as Harper enacted a somewhat tongue-in-cheek version through pre-drawn sketches and various props quickly manipulated on a table and projected through a live feed camera. The field of purple-sprouting broccoli, at first a small single sheet of green felt with one floret of actual purple-sprouting broccoli placed in the centre, became larger and larger as sheets of green felt and florets of purple-sprouting broccoli covered the table and filled the screen. The harvesting got faster and faster as Harper first filled one small toy wagon pulled behind a tractor, but soon needed to use larger and larger toy cargo trucks. As Pettitt spoke of her father and brother reluctantly helping her with the harvest, Harper repeatedly placed and removed cartoon-like paper cut-outs of a farmer into the fields of purple-sprouting broccoli. And when her father showed his annoyance, Harper slapped the farmer cut-out onto the table soon followed by a cut-out 'speech bubble' saying the words that audiences heard Pettitt saying on the other screen. While Pettit's narrative was lively and engaging, the visual storytelling highlighted the enormity of her achievement in spite of the hard work involved in planting and harvesting the crop, in marketing an unfamiliar vegetable to restaurants and supermarkets, and in overcoming hurdles set by her family. Her communication skills were revealed within the story itself, but also in her storytelling skills as she gently ridiculed her overbearing ambition, and the visual storytelling sought to amplify that self-awareness (Fig. 2.6).

The performance ended with all the women appearing on the two screens and moving in and out of prominent positions as though in conversation with each other as they commented on what they thought would happen to farming in the UK after Brexit. Harper moved into the audience. As the lights came up, a woman from the audience rose and moved into the performance area. The audience easily recognised her as one of the interviewees. She then facilitated the live discussion to encourage dialogue about current farming issues raised in the performance and to steer the conversation away from the show itself.

Clearly, *Who's Driving the Tractor?* was a theatre piece rather than a sociological study, so it did not seek to make broad claims about women in UK agriculture. Rather, it sought to uncover what agrarian feminism meant to a small number of women farmers and agricultural experts and to experiment with ways to enable them to tell their stories directly to the public. Sarah Pettitt's miming of putting on layers of clothing to work in

Fig. 2.6 *Who's Driving the Tractor? Conversations with Women in UK Agriculture*. 2018. Performance photograph with Sarah Pettitt telling the story of purple-sprouting broccoli on screen. Performer: Sarah Harper. Photographer: Susan Haedicke

the wet and cold Lincolnshire fields or harvesting her purple-sprouting broccoli faster and faster as demand for it grew made it possible for spectators to feel the hard work viscerally.[37] And Ali Capper's pride in her accomplishment of successfully promoting British beer was reinforced by hand gestures pointing to her, like taking a small fold of her shirt between her thumb and forefinger and pulling it out slightly before patting it back where it had been. While the women offered varied interpretations of agrarian feminism particular to their agricultural roles and personalities, a common thread emerged in their focus on what they each characterised as women's superior skills in communicating, in making connections beyond the farming sector and in listening and responding. They each, again in a variety of ways, emphasised the importance of communication and food citizenship for the health of the industry, and they each displayed pride in their contributions to facilitating those conversations. Charlotte Hollins'

[37] These gestures are visible in the video of *Who's Driving the Tractor?*.

community outreach activities on Fordhall Farm play as important a role in the farm's identity and economic resilience as her brother's care of the farm animals. And Becca Stevenson's work on Five Acre Farm cultivating and harvesting the vegetables is paired equally with her education of the public on the sources of their food through the farm's role in community supported agriculture. Each of the women we interviewed played a significant role in transforming food consumers into food citizens, although none used those words. While there is much we would do differently if we reworked *Who's Driving the Tractor?*, I do feel that we let these women define their own visions of agrarian feminism and social sustainability (Fig. 2.7).

The many storied farmscapes told in words, practices and repetitions of farm tasks by the farmers themselves provide 'insider' stories into these complex landscapes. The artists in this chapter invite the general public into the agricultural world through the narratives of the farmers in their own words and through performed versions of farming tasks. In so doing,

Fig. 2.7 *A Binding Agent*. Pen and ink drawing by Kristen Oshyn

the artists develop the 'experience' of communication as an artistic practice to inform and engage audiences and offer models for establishing a food democracy that depends on changing roles and relationships for both farmer and participating public.[38] The talk-backs that were integral to *Farmscape*, *Map of My Kingdom* and *Who's Driving the Tractor?* and Hollins' storied walk begin to provide audiences with more active roles as the conversations seek to encourage critical reflection about the issues raised. Before looking at the ways in which artists facilitate dialogue in durational projects devised to effect social change, it is important to hear the farmland's side of the story. The next chapter looks at the language of landscapes and how farmscapes in particular can lead to political activism and food citizenship.

Works Cited

AgArts. 2021. https://www.agarts.org/.

Allegue, Ludivine, Simon Jones, Baz Kershaw, and Angela Piccini, eds. 2009. *Practice-as-Research in Performance and Screen*. Basingstoke: Palgrave Macmillan.

Arlander, Annette, Bruce Barton, Melanie Dreyer-Lude, and Ben Spatz, eds. 2018. *Performance as Research: Knowledge, Methods, Impact*. London and New York: Routledge.

Bolton, Susannah. Personal interview. 28 March 2017.

Borda, Sylvia Grace. 2015a. *Farm Tableaux Finland*. http://www.sylviagborda.com/.

Borda, Sylvia Grace. 2015b. 'More about the Processes and Ethics Behind the Development of *Farm Tableaux Finland*'. http://www.sylviagborda.com/.

Borda, Sylvia Grace in collaboration with John M. Lynch. 2015c. *Farm Tableaux Finland Google Street View*. [YouTube] https://www.youtube.com/watch?v=wPqndBnCcxM.

Cannon, Derek. Personal interview. 3 June 2017.

Capper, Ali. Personal interview. 21 April 2017.

Carolan, Michael S. 2008. 'More-than-Representational Knowledge/s of the Countryside: How We Think as Bodies'. *Sociologia Ruralis* 48.4: 408–422.

Collier, Rosemary. Personal interview. 23 March 2017.

Devine, Jenny. 2013. *On Behalf of the Family Farm*. Iowa City, Iowa: University of Iowa Press.

[38] See Food Ethics Council (2017: 19–31) that explores a reimagining of both the agricultural organisational roles (shareholders, producers, retailers, government and NGOs) through a lens of the Citizen mindset and the responsibilities of a citizen public.

'Drama of the Land'. 2015. [Video] Harvest Public Media. https://maryswander.com/media-appearances/.
Drummond, Caroline. Personal interview. 9 May 2017.
Ellison, Tess. Personal interview. 3 June 2017.
Filewod, Alan. 1987. *Collective Encounters: Documentary Theatre in English Canada*. Toronto, Buffalo and London: University of Toronto Press.
Filewod, Alan. 2008. 'Collective Creation: Process, Politics and Poetics'. In *Collective Creation, Collaboration and Devising*. Ed. Bruce Barton. Toronto: Playwrights Canada Press. 1-13.
Food Ethics Council. 2017. 'Food Citizenship: How Thinking About Ourselves Differently Can Change the Food System'. https://www.foodethicscouncil.org/app/uploads/FoodCitizenship2017_web.pdf.
Fordhall Organic Farm. https://www.fordhallfarm.com/.
Gardner, Lynn. 26 March 2016. '*The Shepherd's Life* Review—Gentle Farming Wisdom and Animal Magic'. *The Guardian*. https://www.theguardian.com/stage/2016/mar/29/the-shepherds-life-review-theatre-by-lake-keswick-james-rebanks.
Guyler-Alaniz, Marji. 2020. *FarmHer*. https://farmher.com/.
Haedicke, Susan and Sarah Harper. 2018. *Who's Driving the Tractor? Conversations with Women in UK Agriculture*. https://www.youtube.com/watch?v=Jc-jwkk6p_8.
Hamer, Emma. Personal interview. 12 May 2017.
Haraway, Donna. 2008. *When Species Meet*. Minneapolis, MN: University of Minnesota Press.
Haraway, Donna. 2016. *Staying with the Trouble: Making Kin in the Chthulucene*. Durham, NC and London: Duke University Press.
Hassanein, Neva. 2008. 'Locating Food Democracy: Theoretical and Practical Ingredients'. *Journal of Hunger and Environmental Nutrition* 3. 2–3: 286–308.
Healey, Michael. 1999. *The Drawer Boy*. Toronto: Playwrights Canada Press.
Hellen, Ruth. Personal interview. 3 June 2017.
Hollins, Charlotte. Personal interview. 26 April 2017.
Hollins, Ben and Charlotte. 2008. *The Fight for Fordhall Farm*. Much Wenlock, Shropshire: Ellingham Press.
Holmes, Hannah. Personal interview. 3 June 2017.
Ingold, Tim. 2011a. *Being Alive: Essays on Movement, Knowledge, and Description*. Abingdon, UK: Taylor & Francis Group.
Ingold, Tim. 2011b [2000]. *Perception of the Environment: Essays on Livelihood, Dwelling and Skill*. New York and London: Routledge.
Jones, Ffion, n.d. *Woollying the Boundaries*. Book 1: Fieldwork. Unpublished photographic book.

Jones, Ffion, n.d. *Woollying the Boundaries*. Book 2: Artistic Practice. Unpublished photographic book.

Jones, Ffion. 2014. *Woollying the Boundaries: Perceptions of, and Interventions into, Upland Sheep Farming in Wales*. [PhD Thesis] https://pure.aber.ac.uk/portal/files/10625276/Jones_Ff.pdf.

Jones, Gwendolyn. 2020. 'Women in Agriculture—What They Bring to the Table. Anco Animal Nutrition Competence. https://www.anco.net/women-in-agriculture/.

Kershaw, Baz and Helen Nicholson. 2011. *Research Methods in Theatre and Performance*. Edinburgh: Edinburgh University Press, Ltd.

Little, Jo. 2002. *Gender and Rural Geography: Identity, Sexuality and Power in the Countryside*. London and New York: Prentice Hall.

McCollough, Katelynn. 2014. 'Practical Farmers of Iowa Play Tackles Tricky Farmland Transfers'. *The Gazette*. https://www.hooplanow.com/articles/1140-practical-farmers-of-iowa-play-tackles-tricky-farmland-transfers.

Midmer, Alice. Personal interview. 9 May 2017.

Mulkern, Audra. 2019. *The Female Farmer Project: Documenting the Rise of Women in Agriculture*. http://www.femalefarmerproject.org/.

Nebbe, Charity and Sarah Boden. 2014. 'Women on the Farm: An Evolving Role'. Iowa Public Radio. https://www.iowapublicradio.org/post/women-farm-evolving-role-0.

Nebbe, Charity. 2020. 'Podcast *AgArts from Horse and Buggy Land* Highlights the Arts in Rural Iowa'. https://www.iowapublicradio.org/show/talk-of-iowa/2020-11-23/podcast-agarts-from-horse-and-buggy-land-highlights-the-arts-in-rural-iowa.

Nebbe, Charity and Ben Stanton. 3 May 2018. 'AgArts Promotes Intersection of Agriculture and Art'. Interview with Mary Swander. Iowa Public Radio. https://www.iowapublicradio.org/post/agarts-promotes-intersection-agriculture-and-art#stream/0.

Nelson, Robin. 2013. *Practice as Research in the Arts: Principles, Protocols, Pedagogies, Resistances*. Basingstoke: Palgrave Macmillan.

Norman, Hannah. Personal interview. 20 April 2017.

Nunn, Robert C. 2008. 'The Meeting of Actuality and Theatricality in *The Farm Show*'. In *Collective Creation, Collaboration and Devising*. Ed. Bruce Barton. Toronto: Playwrights Canada Press. 14–28.

Ondaatje, Michael. 1974. [DVD] 'The Clinton Special: A Film about *The Farm Show*'.

O'Neill, Holly, Tim Lewis and Sirin Kale. 16 August 2020. '"We're Stewards of Our Land": The Rise of Female Farmers'. *The Guardian*. https://www.theguardian.com/food/2020/aug/16/we-are-stewards-of-our-land-the-rise-of-female-farmers.

Pettitt, Freida. Personal interview. 4 April 2017.

Pettit, Sarah. Personal interview. 22 March 2017.

Rebanks, James. n.d. 'One Shepherd and his Beloved Herdwick Sheep'. Interview on BBC Radio 4 Book of the Week. http://www.bbc.co.uk/programmes/articles/5mpgT5k3tSNsGBSc0kBD7fg/james-rebanks-one-shepherd-and-his-beloved-herdwick-sheep.

Rebanks, James. 2015. 'People Like us Are Married to the Landscape'. BBC Radio 4, 16 April 2015. https://www.bbc.co.uk/sounds/play/p02p8mg7.

Rebanks, James. 2016. *The Shepherd's Life: A Tale of the Lake District*. UK and USA: Penguin Random House.

Regan, Marion. Personal interview. 1 May 2017.

Roth, Clare and Charity Nebbe. 17 September 2014. '"Buying the Farm" and Buying the Farm'. Interview with Mary Swander. Iowa Public Radio. https://www.iowapublicradio.org/post/buying-farm-and-buying-farm#stream/0.

Sachs, Carolyn E. 1983. *The Invisible Farmers: Women in Agricultural Production*. Totowa, New Jersey: Rowman & Allanheld.

Sachs, Carolyn E. 2018 [1996]. *Gendered Fields: Rural Women, Agriculture and Environment*. New York: Routledge.

Sachs, Carolyn E., Mary E. Barbercheck, Kathryn J. Brasier, Nancy Ellen Kiernan, and Anna Rachel Terman. 2016. *The Rise of Women Farmers and Sustainable Agriculture*. Iowa City: University of Iowa Press.

Slavchevska, Vanya, Susan Kaaria and Sanna Liisa Taivalmaa. 2019. 'The Feminization of Agriculture: Evidence and Implications for Food and Water Security'. In *The Oxford Handbook of Food, Water and Society.* Eds. Tony Allan, Brendan Bromwich, Martin Keulertz and Anthony Colman. Kindle Edition. Oxford: Oxford University Press: Chapter 15: n.p.

Stevenson, Becca. Personal interview. 20 April 2017.

Strom, Jordan and Sylvia Grace Borda. 2020. *Sylvia Grace Borda: Shifting Perspectives*. Exhibition Catalogue. China: Heritage House and Surrey Art Gallery.

Swander, Mary. 2012. *Farmscape: The Changing Environment*. North Liberty, IA: Ice Cube Press.

Swander, May. 2015a. *Map of My Kingdom*. Unpublished manuscript.

Swander, Mary. 2015b. *Map of My Kingdom*. [Trailer]. https://maryswander.com/productions/.

Swander, Mary. Personal interview. 28 December 2016.

Swander, Mary. 2018a. *Field-to-Fork Tales*. Video. https://maryswander.com/productions/.

Swander, Mary. 2018b. *Field-to-Fork Tales*. Video. https://maryswander.com/farm-to-fork-tales-information/.

Swander, Mary. 2020. *AgArts from Horse and Buggy Land*. Podcasts. Season 1, Episode 2: 'Taking Livestock to Market'. https://maryswander.com/podcasts/.

Swander, Mary. 2021. *AgArts from Horse and Buggy Land*. Podcasts. https://maryswander.com/podcasts/.

Theatre Passe Muraille. 1976. *The Farm Show*. Toronto: The Coach House Press.

Tuan, Yi-Fu. 1974,. *Topophilia: A Study of Environmental Perception, Attitudes and Values*. New York: Columbian University Press.

Usmiani, Renate. 2008. 'A Success Story: Theatre Passe Muraille'. In *Collective Creation, Collaboration and Devising*. Ed. Bruce Barton. Toronto: Playwrights Canada Press. 29-37.

Whatmore, Sarah. 1991. *Farming Women: Gender, Work and Family Enterprise*. Basingstoke, UK: Macmillan Academic and Professional, Ltd.

Whitehouse, Andrew. 2012. 'How the Land Should Be: Narrating Progress on Farms in Islay, Scotland'. In *Landscapes Beyond Land: Routes, Aesthetics, Narratives*. Eds. Arnar Árnason, Nicolas Ellison, Jo Vergunst and Andrew Whitehouse. USA: Berghahn Books. 160-177.

Williams, Sally. 'The Fight for Fordhall Farm'. *The Telegraph*. 29 April 2006. https://www.telegraph.co.uk/foodanddrink/3325249/The-fight-for-Fordhall-Farm.html.

World Bank. 2016. 'Feminization of Agriculture in the Context of Rural Transformations: What Is the Evidence?' Report no: ACS520815. http://documents.worldbank.org/curated/en/790991487093210959/pdf/ACS20815-WP-PUBLIC-Feminization-of-AgricultureWorld-BankFAO-FINAL.pdf.

CHAPTER 3

Conversations with Farmscapes: Traces and Echoes

Farmscape Conversations

'The language of landscape is our native language' (Spirn 1998: 15) claims landscape architect Anne Whiston Spirn. She continues, in the evocative opening to *The Language of Landscape*, to interlace landscapes, inhabitants, language and meaning-making:

> Landscape was the original dwelling; humans evolved among plants and animals, under the sky, upon the earth, near water. Everyone carries that legacy in body and mind. Humans touched, saw, heard, smelled, tasted, lived in and shaped landscapes before the species had words to describe what it did. Landscapes were the first human texts, read before the invention of other signs and symbols. … The language of landscapes can be spoken, written, read, and imagined. Speaking and reading landscapes are by-products of living—of moving, mating, eating—and strategies of survival—creating refuge, providing prospect, growing food. To read and write landscape is to learn and teach: to know the world, to express ideas and to influence others. Landscape, as language, makes thought tangible and imagination possible. (Spirn 1998: 15)

Spirn argues that landscape is a language since its topographical features act as words. Just as words acquire enhanced meaning when contextualised in a sentence, 'the meanings of landscape elements (water, for example) are only potential until context shapes them' (Spirn 1998: 15).

© The Author(s), under exclusive license to Springer Nature Switzerland AG 2021

S. C. Haedicke, *Performing Farmscapes*, Performing Landscapes, https://doi.org/10.1007/978-3-030-82434-1_3

Those contexts speak of naturalcultural associations of humans, nonhumans and locations, and landscape's storylines are the tales of human, nonhuman and land encounters. Spirn gives an example of a 'wolf tree' that looks very different in size and growth patterns of the trunk and branches from the other trees in the surrounding forest. Its large branches spreading horizontally and its thick trunk speak of its earlier life as a lone tree in an open agricultural field. Once the field was abandoned and other woodland trees could grow and limit the wolf tree's ability to get enough light horizontally, its branches shifted direction and began to grow vertically. The wolf tree narrates its autobiography in its physicality. Farmscapes do the same through unexpected features visible in the land that suggest human, nonhuman and natural interventions into the physical environment. Lucy Lippard insists that learning to hear the land speak helps us to place ourselves in a larger environmental context. She comments that as we walk the land, 'we can fantasize (or hypothesize) about all the fences, lumps, bumps, furrows, and tracks that cross it. They may be ancient traces or very recent ones, agricultural, industrial or natural, accidental or intentional… One way to find ourselves is to walk the [land], to think about how the land around us is being and has been used' (Lippard 1997: 125). Her insights echo those of John Wylie (2002, 2005) who claims that embodied landscape encounters can foster emergent identities (discussed in Chapter 1). But Lippard's and Wylie's observations remain human-centric. Learning the language of naturalcultural landscapes shares agency as architects of the environment with the land. As Mitch Rose asserts, acknowledging landscape stories confirms that the landscape 'speaks to the people, beckoning them to care' (Rose 2006: 549). Spirn claims that '[l]andscape is loud with dialogues, with storylines that connect a place and its dwellers' (Spirn 1998: 17), although, she admits, it is often hard to understand: 'The language of landscape makes significant details stand out and helps me frame questions, but reading landscape deeply requires local knowledge. On foreign ground, one needs an interpreter' (Spirn, 1998: 4). So how do we learn the language of local farmscapes? (Fig. 3.1).

This chapter explores the various artists' strategies to become 'interpreters' of farmscapes by making the traces on this 'foreign ground' visible to non-farming publics. The benefits to learning the language of farmscapes are multiple ranging from an enhanced appreciation of the local, working, naturalcultural sites to a sharper awareness of the inextricability of human survival from that of the natural world to the formation of a

Fig. 3.1 *From and of.* Pen and ink drawing by Kristen Oshyn

food democracy to protect the land and its active role in the food system. Rebecca Hosking explains in her 2009 documentary, *A Farm for the Future*, how a 'conversation' between the land and nonhumans reveals that current agricultural practices are destroying the environment. She looked into the impact of ploughing on her family farm and found old footage of a field that, in 1981, showed hundreds of birds landing behind the tractor to feast on the exposed insects. In 2008, a photograph of the same field enabled the farmscape to narrate that there was no longer any life in the soil as no birds followed the tractor (Hosking 2009, 2021).

Ffion Jones' *Ode to Perdurance/Awdl Amser*

Karl Benediktsson and Katrín Anna Lund argue that the metaphor of *conversation* offers a way to engage with landscape stories that assumes a 'two-way communicative process' that 'on the human side ... involves both attention and action' (Benediktsson and Lund 2010: 1) and, on the landscape side, relies on topographical features to affirm its ability to '*do* things, ... to make a difference, produce effects' (Bennett 2010: viii).

Ffion Jones' *Ode to Perdurance/Awdl Amser* (2012), one of three pieces created as part of her PhD research, *Woollying the Boundaries: Perceptions of, and Interventions into, Upland Sheep Farming in Wales* (2014),[1] performed a provocative conversation with her family's farm about their shared past and present. Jones describes the catalysing moment that initiated her conversation with the past of Cwmrhaiadr, her family's sheep farm in Wales: early one morning, her father Glynne, who farmed Cwmrhaiadr at the time, handed her 'a crumbling blue bag… bulging with small angular objects only partially visible through the frail, papery plastic' (Jones 2017: 115). Inside were many old notebooks. 'These are *Taid's* diaries, I thought you might like to have a look at them' (Jones 2017: 115), her father explained. The diaries recorded the many jobs her grandfather (*Taid*) did on Cwmrhaiadr for almost forty years. Many of the entries, especially those between 1965 and 1969, referred to building or repairing ordinary but enduring farm objects—stone steps and walls, fences, barndoors—in a specific place on the farm. Jones wondered whether she could find these objects to discover what they were like so many years later and to see how time and the local climate altered the farm work her grandfather had done:

> The relics that *Taid* left behind are distressed, abandoned and derelict when I find them. What was once a brand-new fence is a soggy, quivering entanglement of rusted steel and chewed-up and spat-out wood. In their rotting, a whole host of other life has taken them over. … What does one make with and of the rotting remains of a life, a lifework? (Jones 2017: 116)

In the resulting project, *Ode to Perdurance*, Jones engaged in a conversation with her family's farmscape as she listened to its intensely local stories over the many years of its working life visible in its agricultural traces. She responded to the farm's provocation of 'the rotting remains of a life, a lifework' with artistic practices that established a dialogue across time with the farmscape of the past and that drew attention to the ways in which 'the remains take on a life of their own' (Jones 2017: 116). The working landscape with its local particularities and its interspecies interactions as insects eat the wood or wind and rain transform the stone steps

[1] Chapter 2 discusses one of the other pieces.

and fences (the three modalities of farmscapes discussed in Chapter 1) visibly coalesce in this performance project.

Ode to Perdurance[2] is a series of ten short stop-animated films inspired by the objects on the farm and the farming tasks recorded in her grandfather's agricultural diaries. The opening lines of the series explain that the films:

> have been made with objects that are left in the landscape. The residual traces of Taid's farm work (as evidenced in his diaries) reanimated in a playful manner in order to draw attention to the passage of time and how current farm life and work is constantly drawn into encounters with the past on a daily basis. (Jones 2012, n.p.)

The passage of time appears through a visual storytelling as the objects are transformed. The materialities of the farmscape, detailed in the diaries and the films, offer more than their decaying physical presence as they also hold traces and echoes of the living, working farm left on the land by humans, nonhumans and natural forces alike. And Jones' dialogue with these physical traces acts as a reminder that the farmscape is an ongoing and dynamic partner. Each of the films begins with the date and written text of the entry in the farm diaries in Welsh and English followed by the stop-animated film that shows the object mentioned in the diaries as she found it and the whimsical changes she makes to its appearance. The films enable the farmscape to speak about the past through landscape features created by the deteriorating farm objects and Jones to answer by taking the initial prompt and reworking it into an artistic form. She participates in a conversation with the farm, a conversation between past and present, between her grandfather's farm work and her performance-based work, and in so doing, Jones allows the audiences to *eavesdrop* on this conversation. For Jean-Luc Nancy, this type of attentive listening is linked to discovering a secret. It goes beyond 'hearing', that Nancy equates with understanding the sense of the words or sounds, to a 'listening' that is 'straining toward a possible meaning and consequently one that is not immediately accessible' (Nancy 2007: 6). By encouraging viewers to listen in on her conversation with the farm via her grandfather's farm work and

[2] See Jones (2012), https://www.youtube.com/watch?v=6FeJK_O-HWc. Also see Jones (2017) where she offers a detailed analysis of the creative process of *Ode to Perdurance/Awdl Amser*. See also Jones n.d.1 and Jones n.d.2.

her artwork, Jones suggests that the aesthetics of the landscape is not only in its stunning vistas, but also in its everyday farm work and, in fact, neither should be overlooked. This realisation, Jones argues, may, in turn, help counteract the romanticised visions of farming that often get in the way of tackling pressing rural problems.[3]

In the first film, in response to the entry of 12 January 1966, '*Codi rêls odan steps y dairy/* Put rails up underneath the dairy's steps',[4] the time-lapse footage depicts moss as it seems to grow all over the rotting wooden post and rails visible under a set of stone steps. The passage of time is palpable, and the gradual hiding of the rails by the moss suggests a sense of changing priorities on the farm. The same technique is used in the opposite direction in '*Codi pwt o wal o flaen drysau mawr y ffald/* Built a small wall in front of the yard's big doors' (17 January 1969). Here, a low stone wall is exposed as the moss and ferns that hide it are removed. Jones explains that this process of excavating the wall mimicked her childhood activity of scraping away the soil at the farm's tips to reveal bits of pottery and other remnants from a demolished mansion on the farm (Jones 2017: 117). Here, another voice—that of Jones as a child—enters the conversation. Anecdotal information like this is not offered in the films, but the article's background farmscape stories enrich the playful visual transformations of the objects. I suggest that this article, acting as a companion to the films, adds more than an explanation of creative process as it too enters the conversation between Cwmrhaiadr and Jones and draws attention to a lifelong relationship with the specific place of the lived, working farmscape that has shaped her current experiences. An entry from 16 February 1965 explains '*Codi pyst yn top y paddock/*Put up a post in the top of the paddock'. Here, an old post is gradually covered with rushes from a clump nearby seemingly to suggest the post's ageing as it appears to be growing a long beard. The rushes, for Jones, again brought back memories of childhood games: 'We used to plait them, something my maternal grandfather had once shown us' (Jones 2017: 117). Another reflection on a farm post from 15 February 1968 '*Diwedd y ffens ochr draw I'r afon yn Cwmrhaiadr-Fêch/*Finished the fence on

[3] See Jones (2014).

[4] The translations from the Welsh are slightly different in the films and Jones' article (2017). I am using the ones from the film.

the other side of the river in Cwmrhaiadr-Fêch' shows a rotten fence-post disintegrating bit by bit until nothing is left. Jones explains that her process of breaking down the post was more difficult than expected since 'although the outside was soft and rotten, the core of the wood, or the heartwood, was still very hard' (Jones 2017: 118). Here, the farm work of the past 'argued' with the artist's work of the present. One film, in particular, evokes Jones' childhood memories that can be vicariously experienced while watching the film. In the 29 December 1966 entry, *Taid* had written: '*Trwsio stonder drws Ysgubor*/Fix the Ysgubor door surround'. The Ysgubor was the shed for the sheep dogs, and the film opens with a long list of names of the Cwmrhaiadr sheep dogs on one door and the drawing of a shepherd with a crook on the other, both in chalk. That notation and drawing on the door harked back to the way her father would keep track of the numbers of sheep in various pastures (Jones 2017: 118). During the film, the image and the names are slowly washed away, metaphorically presenting the passage of time as the dogs and her grandfather died over the years and leaving the viewer with a profound sense of loss. These films give evidence of lifelines and storylines from the past, and *Ode to Perdurance* exposes and 'interprets' the shepherding traces left by her grandfather in both his diaries and on the land. It is the persistence of these traces on surfaces of paper and farmlands that provides a key to previous life histories of human-nonhuman-land collaborations in the local, working, naturalcultural farmscape. Tim Ingold describes a trace as a visible mark etched into or built onto a surface by 'continuous movement' (Ingold 2007: 43), like trails created by livestock and farmers or stone walls built to separate pastures. Traces, he argues, suggest links to the past as they are the remains of gestures, activities, and movements and, as such, are dynamic. They seemingly encompass past actions and future possibilities almost as one. Quoting the artist Paul Klee, Ingold explains that such a free-spirited trace is a line that 'goes out for a walk' (Ingold 2007: 73) and invites us to follow the same path, whether on paper or the land.

While the life-worlds depicted in *Ode to Perdurance* represent an homage to Jones' grandfather's work on the farm, in the films, his actual labouring body is never seen. For Jones, the lack of seeing the labour being done in both the films and the diaries actually makes the presence of the labour more urgent: '[t]here is a sense of a body haunting these objects' (Jones 2017: 119). Jones argues that this invisible presence of the labouring body 'draws attention to the messy mortality of both living and material things' (Jones 2017: 119). And her labouring body as filmmaker haunts the objects as well as we watch them transform. In many of the films, the voices of Jones' father and brother, their hammering as they repair the sheep pens and other sounds of farm tasks and animals not only testify to the passing down of local hands-on knowledge from one generation to the next, but also, Jones claims, this soundscape 'suggests a future haunting of this landscape by the ongoing nature of the farm work' (Jones 2017: 120). Their recorded voices resonate from the recent past and echo her Taid's voice as he performed the tasks written in shorthand in his diaries, and reverberate back even further to shepherds working this farm over the centuries. But they also predict future farm work and interspecies relationships. While *Ode to Perdurance* ostensibly seems a relatively simple film project, its sophisticated weaving of the specificity of the local, the often invisible work on the farm and its understated, but ever-present, recognition of the inseparability of nature and culture, evidenced in Jones' conversation with Cwmrhaiadr in the films, attests to its presence as a performed and performing farmscape.

Friches Théâtre Urbain's *Hope is a Wooded Time*

Hope is a Wooded Time (2012–2014), a site-specific, eco-agricultural performance project in Montreuil, just outside Paris, France, sought to make visible the palimpsest of past and present sites of food production—specific centuries-old industrial-scale farming practices and today's foraging practices—that co-exist as traces, echoes and current activities in this neglected woodland farmscape creating a web of place-time narratives. From the seventeenth to nineteenth centuries, much of Montreuil was divided into long narrow plots of land by 600 kilometres of 'living walls': les *Murs à Pêches*. Built of earth, plaster and gypsum, these walls provided heat and humidity to the espaliered peach trees that became a major source of income for the town and supplied Paris with fresh peaches for much of the year. The peaches, each one individually wrapped, were carried in woven willow baskets on the women's heads as they walked each morning from Montreuil to the market in Les Halles in the centre of Paris to sell their fruit. By the mid-twentieth century, many of the plots had become wild, overgrown and full of rubbish, and their walls had fallen into disrepair, but their agricultural traces were left on the land. In 2010, Montreuil launched an initiative that offered grants to site-specific projects, both horticultural and artistic, to respond to the unique aspect of the city's agricultural history found in the *Murs à Pêches*. Sarah Harper, Director of Friches Théâtre Urbain, received artistic management of Parcelle 343, an abandoned woodland site of one hectare on rue Saint-Antoine, for *Hope is a Wooded Time*.[5] Like the remnants of the farm objects on Jones' family farm, the traces of the old peach orchards narrate an autobiography of agricultural work for those who care to listen. The multiple forms of performative response to these remnants in *Hope is a Wooded Time* established naturalcultural multi-voiced conversations between Parcelle 343 and the artists and audiences.

This eco-art project, for which I was dramaturg, drew inspiration from the site's evocative heritage certainly, but also from its contemporary reality. Crisscrossed by the living walls, Parcelle 343 is now a protected woodland site, so trees cannot be cleared, walls demolished, or structures built. However, Parcelle 343 challenges expectations of a beautiful ancient woods since it is overgrown, unkempt and often spongy underfoot; industrial waste fills one corner; pollutants have made the soil toxic in places; and wine bottles, discarded clothing, broken tools, old furniture and human excrement are scattered throughout. And yet, closed in on itself for over forty years, the site became a thriving ecosystem rich in biodiversity. Hidden behind tangled vines and brambles, one can find ruins of the living walls and old wells divulging remarkable centuries-old horticultural and irrigation techniques; figs, quince, chestnuts, hazelnuts and edible greens feeding birds, foxes, mice and human neighbours; mushrooms blooming in abundance under their partner beech, birch, poplar and oak trees in a symbiotic relationship; and even a few very old and gnarled peach trees within the ecosystem of this former agricultural site.

Surrounding this eco-community of living organisms are diverse groups of people. The primary groups of local inhabitants are travellers who squat in neighbouring plots and car parks: Sinti gypsies who have lived in extended family groups in caravans in this area of Montreuil for forty years and a recently-arrived small group of Rom Kalderash, economic migrants from Romania. Other neighbours are long-established French working-class families and recent Russian and North African immigrants. In this diverse community, there are also upwardly mobile professionals who have a keen interest in ecological issues and new agricultural techniques. The many different performative and pedagogic

[5] For additional information, photographs and drawings, see Friches Théâtre Urbain (2012) and n.d. See also, Haedicke and Harper (2015).

activities that constituted *Hope is a Wooded Time* sought to reveal both the woodland's traces from the past as part of a large-scale industrial food production operation and its present-day role as a foraging site for local residents. The project also sought to valorise biodiversity and cultural diversity in this section of the Murs à Pêches by creating a space for conflicting voices, opinions and memories of the neighbours to be heard and exchanged within an artistic context and by asking neighbouring communities to be collaborators in the creative process to interpret this 'foreign ground' (Spirn 1998: 4).

From its inception in 2012, *Hope is a Wooded Time* encouraged multi-layered conversations and encounters, spatial and temporal, between the local place, the artists, Parcelle 343's ethnically diverse neighbours, soil specialists, horticulturalists, local historians, the general public and the nonhuman inhabitants in the wood. In the three-year project, the artists worked with the many diverse groups to learn about the site's place-time narratives and to hear its performative 'voices', both past and present. The project did not seek to change the wood, to gentrify or tame it, but rather to work with it to understand and appreciate its past and present realities and to collect local stories from the people who use the wood in recordings, photographs or filmed portraits. Harper and her community collaborators engaged in multiple artistic and performative activities to start a conversation with the former farmscape site. One key activity was a series of workshops, sometimes with school or community groups, but often with members of the public interested in the *Murs à Pêches*. In some workshop events, small groups of Parcelle 343's neighbours shared their foraging skills honed by years of harvesting the woodland's hidden nuts, fruits and vegetables and their recipes and handed-down knowledge around food preparation of the wood's edible crops. Others mapped the site to reveal its past as a large-scale peach orchard by identifying the location of the walls, wells and old peach trees or highlighted its present-day identity by cataloguing the flora and fauna in this parcel of land. Participants worked on communal creations of ephemeral sculptures with the objects found into the wood or on the construction of eco-objects, like composting toilets. Depending on the time of year, there could be two to three workshops a week. Pedagogic walks drew small numbers of participants to learn about specific woodland inhabitants like the plants paired in symbiotic relationships or to have early morning breakfasts listening to various birdsongs. Working with so many diverse communities gave the project as a whole a very complex dramaturgical structure that often siloed types of activities and groups, and the farmscape stories of Parcelle 343 sometimes got lost. Yet, the project's focused attention on the local specificities of the site, on its working past as a labour-intensive peach orchard and its present as a rich foraging site, and on the parallels between the biodiversity of the wood and cultural diversity its neighbours encouraged an informed and engaged understanding of the *Murs à Pêches*. It also brought varied groups of people from across the food system together to imagine innovative ways forward that, in turn, encouraged food citizenship.

The annual 'ballades artistiques' or self-guided *art-walks* were the largest participatory performative events in *Hope is a Wooded Time*. They sought to instruct diverse participating publics in the language of this unique farmscape. In the last 'ballade artistique' (2014), visitors entering the wood were guided by hand-painted arrows on bits of wood hanging from the trees that took them on a circuitous route through the old peach orchard. Visitors walked past mushrooms growing abundantly under the beech trees and could read texts from an email exchange between Harper and a mushroom expert on their symbiotic relationship that were scientifically accurate, but also veered towards personifying the beech and the mushrooms. Motion-sensor activated recorded sounds of the site itself or stories from the wood's human neighbours pointed out the aspects of Parcelle 343 easily missed by casual observers, for example, rare plants that worked with the land to neutralise the toxins in the soil. Many of the installations in the *Ballade artistique* highlighted past and present hidden agricultural labour. Its past as a productive, labour-intensive peach orchard appeared in staged photographs depicting the harvesting of the peaches projected on some of the still-standing walls, and installations suggesting the hard work of harvesting peaches were created by placing a ladder, a bucket and old farming clothing around remaining, albeit no longer productive, peach trees (Fig. 3.2).

An abandoned and dried-out well also spoke of its agricultural past. When a passer-by peered into it, recordings that suggested echoes of farmers' voices could be heard talking about how the orchard was irrigated with the well's waters or children's voices about how overly curious children fell into its depths. Other installations revealed the wood's contemporary presence as a foraging site. As walkers followed the woodland path, motion-sensor controlled recordings by the site's neighbours told of foraging there and offered recipes for these local foods. Deep in the wood, the audience came across a woodland café with stumps and

Fig. 3.2 Peach tree installation. *Hope is a Wooded Time*. 2012. Photographer: Susan Haedicke

bits of broken wall for stools piled high with food for tasting: mushroom pâtés, quince and fig jams, nut tarts and other titbits made from food foraged in the woods.

By late 2014, the city was no longer able to adequately fund artistic projects, so *Hope is a Wooded Time* drew to a close. In the traces of the orchard farmscape and the realities of the contemporary forage-scape, the land told an unfinished story about past mono-culture productivity and present-day foraging, and the art walks and other communal activities began a conversation with the local site. Learning how to discern the old peach orchards hidden in the woodland site enabled the participants to recognise the stratification of this 'foreign ground' (Spirn 1998: 4) and to learn about centuries-old large-scale agriculture. The project encouraged knowledge exchange about unfamiliar food production methods and sustainable environments and so contributed to the development of informed and engaged food citizens.

Louise Ann Wilson's *The Gathering/Yr Helfa*

Louise Anne Wilson's *The Gathering/Yr Helfa* (2014) enabled her audiences to engage in an 'act of remembrance' through a site-specific walking performance along Watkin Path through Hafod y Llan, a working National Heritage sheep farm on the slopes of Mount Snowdon in Wales. Commissioned by the National Theatre of Wales that aims to produce site-oriented works in Welsh locations that interrogate the complexities and contradictions of the site, *The Gathering* created juxtapositions of past and present, of fact and fiction, to make this familiar site with an abundance of local lore unfamiliar. While the 'gathering' in the title refers to bringing the sheep down from the mountain summer pastures to the lower winter ones, it also suggests a *gathering* of embodied farmscape encounters with narratives embedded in the landscape during this five-hour-long walking performance.[6] For Ingold, landscape is:

> an enduring record of—and testimony to—the lives and works of past generations who have dwelt within it, and in so doing, have left there something of themselves....the landscape tells—or rather *is*—a story...It enfolds the lives and times of predecessors who, over the generations, have moved around in it and played their part in its formation. To perceive the landscape is therefore to carry out an act of remembrance, and remembering is not so much a matter of calling up an internal image, stored in the mind, as of engaging perceptually with an environment that is itself pregnant with the past. (Ingold 2011b: 189)

As participants climbed the mountain on which the sheep graze in the summer months, they could revel in the startling vistas at the same time as the performance made them aware of the hard work both of the shepherding being practised around them as it had for centuries and of mining through the traces on the land from the industrial past of the farm's nineteenth-century slate quarry (Fig. 3.3). The tensions caused by this juxtaposition of beautiful idyllic scenery and dirty, tiring work both from the past and in the present enriched the complexities of the performed and performing farmscape. The idealising penchant, evident in spectator reactions to the stunning views of many of the farmscapes

[6] See Wilson 2021 for photographs of the performance, design elements, maps and research. See also Pitches (2020: 150–153) for an eye-witness account of the walking performance and Wilson (2016) and n.d.

Fig. 3.3 Participants walking on Watkin Path. Louise Ann Wilson's *The Gathering/Yr Helfa*. 2014. Hafod y Llan Farm, Snowdon, Wales. Photographer: Lizzie Coombes

discussed in the book so far, is hard to suppress. I do not think that the recognition of the beauty of farmlands is problematic in itself; it is only so when the iconic scenery becomes the sole identity of the farmscape. Landscape beauty can draw the spectator in and help him or her perceive the storied farmscape of lives and encounters that came before and simultaneously contribute to an emerging sense of self. Seeing and appreciating the beauty, if paired with harsher farming realities, can work in ways like humour in stories of hardship. Humour can create counter-narratives that enable audiences to laugh as they listen to and learn from difficult truths.[7] Like humour, the beautiful vistas can assist audiences to see and hear the farmscape stories they might prefer to ignore, stories that involve exploitation, abuse, danger, pain and death. The landscape beauty can bring the difficult realities of farm work and its low pay and insecurities into sharper

[7] In Chapter 5, I look at this use of humour in El Teatro Campesino's *actos*. See also Paolucci and Richardson (2006).

focus since it presents such a vivid contrast. Storied farmscapes are stories of living and dying, of joy and pain, of abundance and scarcity. Farmscape stories are food stories. Haraway insisted that eating and killing go hand-in-hand: 'eating also means killing, directly or indirectly, and killing well is an obligation akin to eating well' (Haraway 2008: 296). Perhaps the beauty of certain farmscapes can help audiences 'stay with the trouble' (Haraway 2016), and that new knowledge can lead to citizen activism for just and sustainable agriculture. To understand the complexities of farmscapes, it is better to acknowledge and appreciate the beauty, and to place it within contexts of rural challenges like corporate take-overs of small farms (as in Swander's *Farmscape* in Chapter 2), capitalist agri-food systems (Chapter 4) and exploitation of land and labour (Chapters 5 and 6). The beauty matters.

Wilson spent several months over a four-year period on Hafod y Llam, where she experienced the daily practices of sheep farming and then transformed the farm, its current everyday activities and its history into a six-kilometre walking performance up the mountain and back to the farm.[8] This site-specific walk was punctuated by old photographs, choreographies of shepherding and mining, and thematically constructed performance installations that:

> were 'drawn into' the landscape at specific locations chosen for their visually-striking or symbolic features such as a boulder, a waterfall, a derelict shepherd's hut, an industrial scar on the mountainside, and empty house or a lambing barn. (Wilson 2016: 29)

Wilson explained that she saw this farmscape as a way to articulate and reflect on life events and, simultaneously, to map the farmscape's dynamic presence onto the human and sheep lifecycles. For her, *The Gathering* was inspired by the interconnectedness of the stories in the land and the specific histories of the family that shepherded in Snowdonia for generations,[9] and this artistic strategy, I think, draws attention to the farm's naturalcultural modality by highlighting that symbiotic link. Like *The Only Places We Ever Knew*, the walking performance in the previous

[8] *The Gathering* was performed three times in September 2014.

[9] Personal interview with Louise Ann Wilson, 27 July 2018. All references to her stories about the performance come from this interview.

chapter, *The Gathering* did much more than offer historical and contemporary facts as it created a storied walk through place and time.[10] While the performance of *The Gathering* drew these varied stories out of the site by making their traces visible, Wilson also embedded an overarching narrative of 'missing life-events'[11] around fertility and infertility into places on the walk by making the reproductive life-cycle of the ewes a metaphor for human reproduction. These overlapping narratives were evident throughout. Early in the walking performance, the audience passed a derelict shepherd's hut filled with fleeces from sheared sheep. Here, The Woman[12] dressed all in white sings about the joy of pregnancy as she sinks into a large pile of fleeces.[13] A very different reference to fertility could be seen near the half-way mark on the walk up the mountain where the audience could read a poem written for this performance by Gillian Clarke, National Poet of Wales. Painted on a boulder alongside the path, it spoke of the death of the old ram no longer able to tup the ewes[14] and reminds us that illness and death have as significant roles in farmscape stories as health and birth. On the farm, in the final section of the walking performance, a video installation projected onto three shovels showed the annual checking of the ewes for pregnancy and colour-coding of those carrying twins or triplets and marking the 'empty' ewe with black dot. Later, Wilson linked the black dot identifying the barren ewes to women experiencing infertility, particularly in a performance installation in the old farmhouse where The Woman, now in black, lamented infertility as she stands in front of the old kitchen fireplace. The sense of loss is palpable:

> The womb clenches like a heart
> Against the void as they wait,

[10] As part of her creative process, Wilson story-boards these journeys. See the Design section of *The Gathering/Yr Helfa* on Louise Ann Wilson's website for examples of her story boards, https://louiseannwilson.com/work/the-gathering.

[11] Wilson (2016) explores this phrase in detail.

[12] The characters in *The Gathering/Yr Helfa* were all named like this: The Woman, The Boy, The Old Man, etc.

[13] Lyrics and photographs are available on *The Gathering/Yr Helfa* website in the Gallery section, https://www.louiseannwilson.com/work/the-gathering.

[14] Gillian Clarke's poems written for *The Gathering/Yr Helfa* were shared with the author by Louise Ann Wilson. Some of these poems were later published by Carcanet in Gillian Clarke, *Zoology* in 2017.

Weighted for a journey....
Branded barren. Empty.
Hollow on the holy hill,
Without foetus, without future. (Gillian Clarke)[15]

The performative interventions drew attention to the traces on the land as a way to remind the walkers of the farmscape's multiple meanings. The walking performance became an embodied deep map 'formed through the gestural re-enactment of journeys *actually made* to and from places that are already known for their histories of previous coming and goings' (Ingold 2007: 84). Anthropologist Christopher Tilley put those ideas into his own walking experiences:

> to walk is to fuse past with present with future. Walking gathers known past histories, practices and traditions as for the most part, following a path, I am walking where others have walked, in the footsteps of previous generations, the ancestors....Walking a landscape is thus to gather together through my body its weathers, its topographies, its people, histories, traditions, identities. The walk gathers itself through my own body to create my own identity. (Tilley 2012: 18)

As participating walkers followed the footsteps of shepherds, sheep and sheepdogs on the farm, they physically experienced a multi-layered narrative trajectory through time that did not create a single storyline but rather wove together meaning, identity, belonging, imagination, history and multiple life-stories circulating in the landscape. This complex web animated an embodied conversation between a walker and the farmscape that, in turn, contributed to his or her emerging sense of self, as Tilley above observed.

For each of the three performances in September 2014, an audience of two hundred gathered in a field near the farm. They were separated into ten groups, each identified by a flag designed to match the ear-markings distinguishing the various flocks of sheep. The flags were white (colour of the wool) and red (blood), a colour palette repeated in the costumes of the performers and musicians. At the start of the performance, the 'flocks' of spectators were gathered by The Boy who stopped on the path just above the field where the audience waited, blew a conch

[15] The whole poem is available on *The Gathering/Yr Helfa* website in the Gallery section.

shell (inspired by an old photograph probably of the farm cook calling the workers to dinner) to call the participating walkers to follow him and told them that in September all sheep are gathered from the mountain. This character of The Boy was based on an old photograph from the farm's archives of Prys Williams, a boy in the family who lived and worked the farm decades earlier.[16] He was the one character in the production who was based on an actual member of the farm family, so to reflect that identity, his texts throughout the performance focused on farming facts. As each group passed through the gate on its way up the mountain, the walkers were counted like a flock of sheep passing through the farm gate. For me, this metaphoric transposition of spectator to sheep not only gave instructions to move together like a herd, but also to sense the bond—the 'hefting'—of the sheep to the mountain, and it seemed to enhance the role of audience. The walkers passed rubber gloves filled with water hanging from tree branches like udders suggesting the birth and feeding of the spring lambs. In a choreographed moment, The Boy cavorted like a young lamb in the clearing near the bridge and then transformed into a shepherd inspecting the sheep to determine their age, repeating 'milk teeth, two teeth, four teeth, six teeth, broken teeth'. He embodied the interconnectedness of sheep and farmer as his choreography performed the naturalcultural farmscape. As the audience climbed, they could see archival photographs of the sheep farm depicting several generations of the family 'hefted' to this land. At one point on the climb, the audience could look down the mountain to the farm below, and in the distance, they could hear a solo trumpet playing simple notes echoing the bleating of a ewe. Wilson explained that during her research, she was by herself in a field used to gather all the lambs about to be separated from their mothers at the end of the summer when she saw a lone ewe bleating. The trumpet for Wilson echoed the plaintive bleating of the ewe, perhaps calling for her lamb, and became a leitmotif as the trumpet sound travelled up the mountain with the walkers (Personal interview, 27 July 2018). The musical notes, passed from one trumpeter to the next, suggested more-than-human conversations that recurred throughout the performance. The trumpets, the hanging rubber glove udders and the choreographies of The Boy all had a metaphoric presence linking the real and the imagined.

[16] See Fig. 3.5 that includes The Boy.

The farm's mining traces appeared moments later as the audience passed The Tramway Walker, dressed in red, who, at this moment, was in a harness that enabled her to be perpendicular to the rock face she climbed and descended over and over. This repetitive activity imitated the tram that climbed and descended the mountain carrying the slate from the quarry that was located further up the mountain (Fig. 3.4). In the distance, a line of red wool carpet marking the steep path that the tram followed to bring slate down the mountain could be seen. During the performance, The Tramway Walker always followed the tram tracks, so her presence made the tramway visible and, like The Boy, embodied the naturalcultural farmscape. Later, the audience saw her carrying two red flags as she crossed a bridge over the river, simulating a walker who usually preceded the tram engine with red flags warning that the tram was coming (Personal interview, 27 July 2018). Later still, the audience saw her, dressed in white, climbing the red carpet. The two women, The Tramway Walker and The Woman in white in front of the derelict shepherds' hut made visible contrasting working landscapes of mining and sheep farming that the audience could also discern in the actual

Fig. 3.4 'Tramway Walker' ascending the 'Tramway Incline'. Louise Ann Wilson's *The Gathering/Yr Helfa*. 2014. Hafod y Llan Farm, Snowdon, Wales. Performer: Kate Lawrence. Photographer: Lizzie Coombes

present-day landscape: traces of the tramway tracks crossed the Watkin Path, abandoned small stone shepherds' huts dotted the landscape and, in the distance, the actual shepherds and their sheepdogs started to gather the sheep and herd them to their winter pastures.

Part way up the mountain, the ten 'flocks' of walkers were divided into two groups. One followed Watkin Path past a long poem honouring the mountain's past storylines painted on the rock cliff:

> The mountain has secrets,
> Tunnels in the seep and drip of the dark,
> Into the stone womb, under roots of trees,
> Past wheels, pulleys, chains, trucks locked
> In their pollens of rust.
>
> Deep as the Ordovician, old workings,
> Mine-shafts, mullock heaps, piles of slag,
> Abandoned two centuries back, river-stones
> Stained blue with copper, copper-iron's gold,
> Sulphates form the mountain's heart. (Gillian Clarke) (Fig. 3.5)

Fig. 3.5 'This Mountain Has Secrets' fissured-rock installation and The Boy. Louise Ann Wilson's *The Gathering/Yr Helfa*. 2014. Hafod y Llan Farm, Snowdon, Wales. Poem by Gillian Clarke. Performer: Meilir Rhys Williams. Photographer: Lizzie Coombs

Fig. 3.6 Storyboard of upper Amphitheatre with Shepherds and Band by Louise Ann Wilson. *The Gathering/Yr Helfa*. 2014

The other group followed the tram tracks. At the highest point in the climb, in the old Cwm Llan slate quarry buildings, the walkers re-converged to witness Clarke's poem cycle,[17] The trumpets played an important role at the start of this performance-within-the-walking-performance to gather all the walkers. The musicians each played a single note from various places on the mountainside above the quarry buildings. These varied notes gradually came together to create a tune as the musicians moved towards the quarry buildings (Fig. 3.6). For me, that gathering of the notes also foreshadowed the actual gathering of the sheep later in the performance.

Clarke's performed poem cycle not only detailed the farmscape's annual cycle month-by-month, but also performed the farmscape conversations. The Old Man personified the mountain; The Shepherds (actors,

[17] Much of the text can be found on website, *The Gathering/Yr Helfa* website, https://louiseannwilson.com/work/the-gathering, in the Gallery section. Wilson shared the entire script with the author.

not actual shepherds) described shepherding tasks; The Woman expressed sentiments around biological motherhood; and The Boy spoke the words of Arwyn Owen, one of the shepherds at Hafod y Llan, transformed into 'poems' by Wilson to lyrically describe seasonal sheep farming activities. In 'September', for example, descriptions of the gathering juxtaposed the narrative of the shepherds with that of the mountain. The Shepherds explained:

> The dogs streak to the slope.
> A single ewe on the track.
> Then the gathering flock
> a rosary of sheep, strands
> threading their way from ridge and crag. (Gillian Clarke)

The Old Man, in contrast, offered the mountain's perspective:

> Dogs and sheep in the dance of the slopes,
> the mountain deep in animal mind,
> hefting the weight, treading the shape of it,
> the knowledge, *cynefin*
> in their bones, nerves, blood. (Gillian Clarke)

Or in 'November' when the ewes are mated with the ram, The Boy expounded:

> Sheep are seasonal breeders.
> Ewes ovulate with fading light.
> On 1st November:
> Ewes 'have' rams.
> Ewes 'take' rams,
> Rams 'serve' ewes,
> Rams 'tup' ewes,
> Rams 'run' with ewes. (Louise Ann Wilson)

However, The Woman, embodying the ewe, approached the same event emotively:

> In reducing light,
> sleeping hormones wake
> in her brain's dark chamber,

and she is ready, restless again
for the scent of the ram.

And The Old Man spoke for the mountain:

Thunder in the mountain,
thunder in the stones,
old thunder in the mine-shafts,
in the blood-stream, in the bones,
in the caves where rivers start,
in the womb where life begins
like a match struck in the dark. (Gillian Clarke)

The poem cycle performed the farmscape's annual conversation.

As the audience began the descent to the farm, the shepherds and sheep dogs accompanied them along the path for several minutes, 'herding' them to the farm. And soon after, the actual 'gathering' of the sheep came into view. While the gathering of the sheep from the summer pastures high in the mountains was a thrilling dramatic moment for the predominantly non-farming audience as the sheep ran down the mountain guided by the sheep dogs and followed by the shepherds, this transhumance represented only a small moment in the complexity of the performance as a whole, just as the gathering of the sheep is an important, but brief, event in the annual life-cycle of the farm. The audience followed the sheep back to the farm where many installations illustrated the work of a sheep farm. The actual shepherds showed their skills shearing sheep, and the returning sheep were 'checked-in' as each ewe and lamb was counted like the spectators at the start of the performance. In the lambing barn, a triptych of films showed how the shepherds encouraged a ewe whose lamb was stillborn to adopt another ewe's lamb, often one of twins, by skinning the dead lamb, placing its fleece on the live lamb and encouraging the ewe to suckle the lamb (Fig. 3.7).

Installations, like two hundred farm objects (one for each 'broken mouthed' sheep brought down from the mountain), were exhibited in the farm yard. The work of the farm was also highlighted in archival photographs of generations of the farm's family held by school children from the Beddgelert Primary School. The walking performance ended in the lambing field as the musicians again played 'sheep' notes that gathered into a tune and gathered the walkers to witness the bow taken by the shepherds and their dogs.

Fig. 3.7 Storyboard of lambing barn installation by Louise Ann Wilson. *The Gathering/Yr Helfa*. 2014

The dramaturgical complexity of *The Gathering* as a creative process and performance product reflected the entangled storylines of the farmscape's humans, nonhumans and land that narrated the farm's local specificities, revealed the histories of its working life and issued a 'call to care' grounded in caring for ourselves, others and the dynamic world we inhabit (Rose 2006: 542). The performing of these multiple farmscape stories made contradictory processes and possibilities visible as *The Gathering* juxtaposed oppositions of iconic rural scenery and often invisible, dirty and difficult farm labour and contrasted poetic farm imaginings and more disturbing farm realities. But the performing of this farmscape also demanded that we hold onto the apparent contradictions simultaneously and recognise them as indicative of the farmscape's complexity as an ongoing process. By locating fertility and infertility, life and death, in the farmscape alongside its startling beauty, Wilson encouraged her audiences to engage in dialogue and 'becoming-with' the surrounding worlding farmscape. As Haraway argued:

thinking people must learn to grieve-with. ...Grief is a path to understanding entangled shared living and dying; human beings must grieve with, because we are in and of this fabric of undoing. Without sustained remembrance, we cannot learn to live with ghosts and so cannot think. (Haraway 2016: 39)

Mike Pearson's *Carrlands*

Mike Pearson described the agricultural landscape in north Lincolnshire in England as 'simple, flat, often deserted—without moments of conventional scenic heritage: [landscape] that does not easily reveal itself; that lacks those monumental features which frequently orientate our gaze; where the land seems to draw back from the viewer; where few Xs mark the spot' (Pearson 2011: 282), and yet he sought to 'share his enthusiasm' (Pearson 2011: 282) for this place where he spent his early childhood and to reveal its complex local, working and naturalcultural narratives recognisable in its traces and echoes. *Carrlands* (2007) provides an ongoing opportunity to get to know and to immerse oneself (actually or imaginatively) in this place through oral stories, music and sounds that grow out of the landscape into a series of multi-layered, self-guided audio-walks that one can listen to either while meandering in the Ancholme River Valley or sitting at home looking at the photographs on the *Carrlands* website.[18] Each audio-walk focuses on one specific area of the valley, Snitterby Carrs, Hibaldstow Carrs and Horkstow Carrs, as it narrates details about the history, agricultural practices, flora and fauna and place-based anecdotes and myths in four different thematic (and rhythmic) movements. These audio recordings begin to teach the listener the 'language' (Spirn 1998) of this farmscape through a growing awareness of the traces and echoes left on/in the land as humans transformed it from boggy, scrubby fens to fertile agricultural land. The audio-walks

[18] All the *Carrlands*' audio walks are available on the *Carrlands* website, http://www.carrlands.org.uk, so all subsequent references are from this website. Each is identified by the name of the walk, and the number and name of the movement in that walk, i.e. 'Snitterby Carrs 1: Myth'. *Carrlands* was an Arts and Humanities Research Council (AHRC)-funded project as part of the five-year Landscape and Environment strategic programme to interrogate the cultural significance of landscapes and understand how the world is 'imagined, experienced, designed and managed'. https://ahrc.ukri.org/research/fundedthemesandprogrammes/pastinitiatives/strategicprogrammes/landscapeandenvironment/. See also Pearson n.d., 2006 and 2010.

poetically place a number of Xs to 'mark the spots' in these storied farmscapes. Using performative spoken word, sound effects and music, Pearson encourages the listener to experience or imagine three locations in the seemingly bland landscape, to understand the work that went into creating them and to appreciate their local specificity in their stories of the inhabitants and the land and, in so doing, to become stewards of the unremarkable land alongside the farmers past and present. The tensions between the stunning vistas and the working farmscapes so evident in *The Gathering/Yr Helfa* take a different form here. In *Carrlands*, the 'vistas' appear in the lyrical texts narrating legends of local places, in folktales of local traditions and memorable events and in an evocative musical score sometimes on its own, sometimes playing under the texts that suggest a portentous 'beauty' in the farmscapes. These 'vistas' are juxtaposed with information about the agricultural work that reminds the listeners that these farmscapes are composed of the lives of ordinary people living ordinary lives and that farmlands represent their hard work of food production. Geographer Hayden Lorimer calls narratives like these 'small stories', mundane and unexceptional certainly, but rich in insider knowledge of the specificities of a particular local, working, naturalcultural farmscape (Lorimer 2003: 197–217).[19] *Carrlands* follows this localist perspective of small stories—this form of chorography. Denis Cosgrove explains that chorography focuses on the specificity of particular locations 'to understand and represent the unique character of individual places' (Cosgrove 2004: 59, discussed in Chapter 1). Drawing on chorography, Pearson sought to uncover 'the intimate connection between personal biography and the biography of the landscape, between social identities and a sense of place and through concatenations of local observation and critical academic discourse' (Pearson 2011: 283) in this seemingly commonplace landscape. The lyrical texts, sounds and music entice the listener to hear and sense the complexities of these farmscapes, a web of physical, remembered, felt and imagined places, agricultural work, somatic experiences and interactive encounters. The valley becomes a 'landscape then for doing, feeling and contemplating as much as for looking, and for performance that avoids either pointing to or pointing out' (Pearson 2010: 82). Pearson seems to suggest in one of the audio-walks that it is the very ordinariness of agricultural Lincolnshire that enables the listener

[19] See also Lorimer (2005).

to project his or her own understandings and emotions onto the farmscape and to locate a sense of self in its imagined topography that is familiar in its blandness: ‘a place to think about landscape, not so freighted with notions of what proper knowledge of landscape is. Maybe as much about what you bring, what you project onto this seeming emptiness, as what you read from it, feelings of disquiet, loneliness, apprehension maybe’ (‘Hibaldstow Carrs 1: Air’). But Pearson does more than reveal connections; he enables the spectator to bring the farmscape to life as the listener is lured into this sensual storied world.

This listening-learning dialogue between the listener, Pearson’s voice in the recordings and the farmscapes bears a family resemblance to Jacques Rancière’s concept of the emancipated spectator. An emancipated spectator does not just receive the artwork as a gift, but rather tests its premises and solutions. The artwork, Rancière argues, remains separate from the spectator and the artist but links the two: what he called a ‘third thing’ (Rancière 2009: 14). It is through this ‘third thing’ offered by one and *translated* by the other, but belonging to neither, that emancipation, a new reflective understanding of the world, can be achieved. A spectator becomes emancipated as he or she gains critical awareness by translating the ‘third thing’ into his or her own experience, by linking it to what he or she already knows and, through that association, creating new knowledge. For Ranciere, emancipated spectators play ‘the role of active interpreters… [and] develop their own translation in order to appropriate the “story” and make it their own story. An emancipated spectator is a community of narrators and translators’ (Rancière 2009: 22). I do not want to push this parallel too far, but Pearson’s reliance on the listener to translate a received audio recording (the artwork) into his or her own lived experience can create a critical awareness not only of the impact of transforming the fens into large-scale arable farmlands on the lives of those who lived and worked there, but also of the importance of the need to care for the land of these large-scale industrial farms for our own survival, and, in so doing, it encourages a willingness to participate in devising a better agri-food system.

The audio-walks do not determine a specific path for the onsite participant with landmarks/traces that correspond clearly to the narrated text, so rather than a guided tour, they create a version of ‘deep maps’ that ‘record and represent the grain and patina of a particular place through juxtapositions and interpenetrations of the historical and contemporary, the political and the cultural, the factual and the fictional, the academic and the discursive... in full cognizance of local and personal knowledge' (Pearson 2011: 281).[20] *Carrlands* offers experiences akin to what Ingold calls 'wayfaring': a constant movement through the world (life's experiences) that focuses on the journey and active engagement with the surroundings rather than on an arrival at a destination. Ingold unpacks what wayfaring means through an 'experiment' with a hand-drawn wavy line that remains 'as a trace of your manual gesture. In the memorable phrase of the painter Paul Klee, your line has gone for a walk' (Ingold 2011a: 150). He asks the reader who drew the initial line to create a similar line through a series of dots that is no longer the trace of a gesture since '[a]ll the energy, all the movement is focused down on the point [of the pencil making the dot], almost as though you were drilling a hole' (Ingold 2011a: 150). It is 'a series of appointments rather than a walk' (Ingold 2007: 73). Reconnecting those dots, he explains, does not recreate the first wavy line, but rather a halting, jerky movement from one spot to the next: 'no longer the *trace of a gesture* but an assembly of *point-to-point connectors*' (Ingold 2007: 74–75, italics in original). The original wavy line, for Ingold, depicts the notion of wayfaring as opposed to a system of transport from one destination to the next. The wayfarer is movement, and the act of 'travel' is a metaphor for passage through the life of an individual, a community, an event.

Ingold asserts a correspondence between a wayfarer and a storyteller as both follow a path 'from place to place—or topic to topic' (Ingold 2007: 91) that one remembers as one goes along. The parallels of wayfaring and storytelling illuminate Pearson's audio-walks of movement, exploration and discovery through time and place as *Carrlands* creates a mode of travelling through the farmscape for his listeners that encourages them to follow leads, suggestions and intuitions to discover a path. This approach does not provide answers but rather suggests lines of inquiry to follow. 'Path-making ... does not so much add another figurative layer to the ground surface as weave another strand of movement *into* it' (Ingold 2015: 61). The listener's way through the various farmscapes in *Carrlands* is evocative and illusive and, like Ingold's wayfaring inhabitants of the world, the listener participates 'in the very process of the world's [or in this case, the farmscape's] continual coming into being' (Ingold

[20] See also Pearson (2006) and Pearson and Shanks (2001).

2007: 81). This acquisition, even construction, of new knowledge of food production constitutes an early stage in becoming a food citizen.

The form of an audio-walk[21] privileges the aural experience over a spectatorial one as a way to *feel* the landscape, to experience it sensorily. The waking up of the listener's imagination through sound was a key tactic in *Carrlands*, and Pearson frequently challenged his listeners to imagine the agricultural landscape from the air ('Hibaldstow Carrs 1: Air'), to picture it in the past (Hibaldstow Carrs 2: Water) or to populate its empty vastness with animals (Horkstow Carrs 3: History). The first audio-walk for Snitterby Carrs prepares the participant for these imaginative thought experiments quite dramatically as it locates the listener in the recognisable present-day pastoral site: 'You may have chosen a spring day, the sun warm on your face, but not yet hot enough to distort the scene with haze and distant shimmering, soil, crops, sky all in distant hues' (Snitterby Carrs 1: Myth'), but it then abruptly pulls him or her into a scary past of threats and superstitions:

> But imagine this. Imagine this as a landscape of darkness, primeval, diluvian, before the dykes were made and the river bed changed, when the carrs were *nobbut bog-lands, an' full o' watter-holes*, when this was a land of great meres of black water and creeping trickles of green water—*squishy mools as'd suck owt in*, a wild desolate dreary marsh full of strange sights and sounds. ('Snitterby Carrs 1: Myth')[22]

Even before the words of this thought experiment begin however, Pearson and musicians John Hardy and Hugh Fowler start to train the listener's ears with close to three minutes of sound—spooky, distorted, alive. One can hear musical notes of reed and brass instruments that seem to morph into bird shrieks or animal howls or harp strings that evoke water and echoes. The eerie sounds are dynamic and pulsating, and they continue almost non-stop throughout this first movement. The sounds, I think, begin to teach the participants *how* to listen deeply to the farmscapes of the Ancholme River Valley to acquire a deeper, somatic knowledge of the landscape. The thought experiment seamlessly shifts

[21] See also Myers (2011).

[22] This case study will include many quotations from the recordings so that the reader can experience the multi-layered interlacings of fact and fiction, of 'vistas' and hard agricultural work, in the texts themselves rather than in more academic descriptions.

into a detailed legend of deep listening and interconnectedness between the early local inhabitants and Tiddy Mun, the Old Man of the Fens before they were drained, an interconnectedness based on affective dialogical listening between inhabitants and landscape. While this story of local superstitions reveals how resonant listening[23] helped the early inhabitants have an embodied conversation with their landscape, personified in Tiddy Mun, it also offers a lesson of listening for the participants in *Carrlands* audio-walks. Pearson evocatively narrates that the inhabitants believed Tiddy Mun lived deep in the water, 'emerging only when the mists rose'. They could barely see him, but were aware of his presence through sound: 'the *limp-elty lob-elty*' of his gait, 'the sound of running water and a sigh of the wind and laughter like a pyewipe's screech. And when they heard the skirl of his passing, they'd pull closer to the fire' ('Snitterby Carrs 1: Myth'). In spite of fear of Tiddy Mun, they would seek his help in flood years, calling '*Tiddy Mun wi-out a name. The waters thruff*'. When they heard the pyewipe's screech, they knew he had answered their cry. The legend continues with Tiddy Mun's curse of pestilence in the land once the fens were drained, and the inhabitants' desire to give the water back.

> So at New Moon, they met at the cross dyke, each carrying a scoop of fresh water. And as darkness fell, looking out over the new river Ancholme, in quavering voices they called: *Tiddy Mun wi-out a name. Here's water for thee. Take tha' spell undone*, whilst pouring out their water—*splash sploppert.* Standing together, trembling, listening, nothing but unnatural stillness. Then suddenly there arose the most awful bawling and whimpering all around them, like a host of little crying babies.... It was the dead children. And they called to Tiddy Mun to let their children live and grow strong. And the innocents in the darklins wailed, trying to reach their mothers' breasts. Women said afterwards that tiny hands had touched them and cold lips kissed them and soft wings fluttered all around them as they endured the woeful howling. Then all at once there was stillness again, and they could hear the water lapping at their feet and a dog yapping in the garth. And then, soft and tender, from the river itself, the pyewipe's screech. It was the Old Man, kind and brooding. ('Snitterby Carrs 1: Myth')[24]

[23] See Heddon (2017) whose concept of 'entangled listening' shares family resemblances to this resonant listening.

[24] Mike Pearson told the author that all the italicised words in the quotations about Tiddy Mun from 'Snitterby Carrs 1: Myth' are originally from three works by M.C. Balfour, all published in 1891: 'Legend of the Cars'. *Folklore* 2,2: 145–170; 'Legends of

Legend has it that Tiddy Mun lifted the pestilence, and for decades, the inhabitants replenished the waters. The story not only evocatively narrates the importance of conversation and careful listening beyond human interaction for survival of these early residents, but, I think, it also resonates for today's listeners who may well have forgotten how to listen to the land as it speaks through its traces and echoes.

The local storied farmscapes in *Carrlands* reveal a past visible in the present in narratives that interweave historical fact, myth, topography, environment and biography in ways similar to the other case studies in this chapter. Understanding the farmscape in this way confirms that its meaning is not in its iconic scenery that can be enjoyed from a position of detached observer. Rather, as we listen, we understand that its power is in the local place-stories and life-stories, in layers of the factual past and present that are as vivid as the myths, and that immersion in the life world of the farmscapes can promote care for the world as its health is linked to our own. '[T]his is no *tabula rasa*, no blank canvas', Pearson explains, 'it is marked at the most intimate of scales—by the benchmark on the bridge, the poignant signs that warn fishermen of the perils of casting near overhead cables, fishermen who no longer come on coach trips from Sheffield, by the burned rubber of joy-riders revving on the bridges, by your boot prints in the mud' ('Hibaldstow Carrs 1: Air'). He weaves a tale of traces so vivid that they became 'visible' to the listener:

> But here and there in the fields long unploughed, close to the bridge, kept for cattle brought down in the summer as soon as the grass was long enough, though it gave them diahorrea, taken back up into the crew yards after one or two frosts, are the marks of former occupations, the odd shadow, the kinds of detail photographed from his biplane in the 1920s by OGS Crawford as he flew north up the Roman Ermine Street, the medieval lumps and bumps at 'Nookings' that local people so want to be the abbey of Saint Hybald. Dig a hole, build a bank, and its trace will survive, the marks of both long-term endeavours and short-term initiatives, schemes, events, accidents, appearing and disappearing over different times scales. ('Hibaldstow Carrs 1: Air')

The audio-walk around Horkstow Carrs also reveals traces left by previous inhabitants over the centuries—stone axes, bronze objects, rapiers and

the Lincolnshire Cars, Part II'. *Folklore* 2,3: 257–283; and 'Legend of the Lincolnshire Cars, Part III'. *Folklore* 2,4: 401–418. They have recently been published in Balfour 2010.

swords, an old log boat, the old sluices and the cuts in the land to enable drainage. These naturalcultural traces are visible in the farmscape when one can read its features.

Pearson suggests that the echoes still reverberate as well:

> in close up, you may encounter remnants, or at least whispers, of older landscapes, of that which has long been here: the rustle of *phragmites* reeds, the plop of a water vole, the call of a lapwing, biding their time perhaps, for the moment when draining this place begins to seem like a losing battle and the electronic sensor on the sluice is switched off. ('Horkstow Carrs 4: Drainage')

The echoes can often be heard in the dramaturgy of the audio-walks as well, for example, as he narrates the story of Percy Grainger collecting old Lincolnshire folk songs, the singers' voices from the old recordings echo in the background ('Horkstow Carrs 3: History') or a cacophony of voices from past centuries can be heard under the story of the many people who passed through Hibaldstow: 'gangs of Victorian women and children potato picking, the Irish lifting sugar beet, the German and Italian prisoners of war from Pingley Camp in Brigg ditching and dyking' ('Hibaldstow Carrs 4: Earth'). This technique of voices or other sounds, like flowing water or bird cries, under the text is complicated even further in the narrative on Ancholme Inland Drainage Board that is responsible for drainage in the area. The rules of conduct that, when recited, could be quite deadly, come alive in the audio-walk, I think, as voice overlays voice reading them. The rules echo in the aural space seemingly increasing in authority until one voice dominates with 'Or you will be breaking the law' ('Hibaldstow Carrs 2: Water'). In another audio-walk, Pearson goes beyond heard sounds as he also reveals the power of silence as a *sound*: 'But in this field or that, close to the river, in this great bowl of accumulation, occasionally you may sense the silence of the deeply blanketed history beneath your feet, hidden, unavailable, but still potent, glimpsed in those remarkable moments when it returns to us' ('Horkstow Carrs 4: Drainage'). This silence is so profound in this vast landscape that he warns, 'Shout here and there will be no echo. Shout here and your voice will travel. Shout here and few will hear you' ('Snitterby Carrs 2: Past').

In *Carrlands*, the audio-walks perform the farmscapes through imaginative engaged encounters between the listener and the site. The nuanced and multi-levelled recordings offer performance experiences 'from which performers are *absent* but within which the audience member plays an active and generative role in the meaning creation, as a *participant*' (Pearson n.d.: 2, Italics in original). Each performance of the *Carrlands* farmscapes is unique to each listener-performer since he or she translates the experience of the evocative, fluid and unrestrained recordings into his or her own experience and makes it his or her own story. That translation, in turn, gives the listener a greater critical awareness of the complex farmscapes of the Ancholme River Valley and can foster a form of Rancière's emancipated spectatorship, or, as is argued throughout this book, a sense of food citizenship, an awareness of the need for consumers to become knowledgeable, engaged and pro-active citizens in agri-food systems. Listening to the audio-walks in the Lincolnshire fields or from a desk chair while looking at the photographs of the sites demands action, and the dialogue between listener-performer and farmscape *appears* the farmscape. 'To *appear things*… is tantamount to imagining them. To imagine something is to appear it, to assist in its gestation and to attend to its birth' (Ingold 2015: 140, italics in original). *Carrlands* enables the listener to imagine and 'remember' these farmscapes through time as Pearson lets us hear and see this layered or vertical landscape, experientially if walking and imaginatively if sitting off-site. For me, being the listener of the vivid aural text as the sole performer/co-creator away from the actual geographical site enabled me to enter the stories, to be immersed in the life world that could shift from past to present, from being air-born to water-logged, in a second. I felt empowered to create my story nested within Pearson narrative. Walking the land, however, I got bogged down in its materialities. Unlike the other case studies in this chapter, paradoxically, the language of the farmscape presented in its traces was most audible away from the land itself.

The case studies in this chapter have explored how the artists encourage the general public to become informed about and engaged in complex farmscapes by detailing them through three modalities of local farmscapes, working farmscapes and naturalcultural farmscapes and by revealing the benefits of learning how to communicate with the land. While Chapter 2 focused on performed conversations with farmers, this chapter has proposed conversations with farmscapes, possible by learning the *language* of farmscapes understood through its topographical features, though traces and echoes made visible and audible by the artists. Hearing the land, Mitch Rose argues, strengthens a commitment to sustaining its health for mutual benefit (Rose 2006: 549). Listening to the land, this chapter claims, not only contributes to an awareness of the urgency of becoming a steward of the land, but also leads to a recognition of one's emerging subjectivity dependent on the interlacings of farmscape and self. This in-depth local insider knowledge gleaned from a conversation with the land itself and from the embodied relationship with the land as a partner contributes a sense of urgency to improve the current food system to protect our world. Chapter 4 will look at case studies that prioritise creating networks, participating in decision-making and understanding how the food system works, and it will explore how the artists foreground dialogue and problem-solving as a part of the artwork, and in so doing, reinforce food citizenship.

Works Cited

Balfour, Mrs. M. C. 2010. *Legends of Lincolnshire Cars*. UK: Read Books Ltd.

Benediktsson, Karl and Katrín Anna Lund. 2010. *Conversations with Landscape*. Farnham, Surry, England and Burlington, VT: Ashgate.

Bennett, Jane. 2010. *Vibrant Matter: A Political Ecology of Things*. Durham and London: Duke University Press.

Cosgrove, Denis. 2004. 'Landscape and Landschaft'. *German Historical Institute Bulletin* 35: 57–71.

Friches Théâtre Urbain. 2012. *L'Espwar est un temps boisé* [Hope is a Wooded Time]. https://www.friches.fr/projets/l-espwar-est-un-temps-boise.

Friches Théâtre Urbain. n.d. *L'Espwar est un temps boisé* [Hope is a Wooded Time]. http://parcelle343.hautetfort.com/about.html.

Haedicke, Susan and Sarah Harper. 2015. '*Hope is a Wooded Time*: An Eco-Performance of Biodiversity in Discarded Geographic and Social Space'. *Performing Ethos* 5.1–2: 101–118.

Haraway, Donna. 2008. *When Species Meet*. Minneapolis, MN: University of Minnesota Press.

Haraway, Donna. 2016. *Staying with the Trouble: Making Kin in the Chthulucene*. Durham, NC and London: Duke University Press.

Heddon, Deirdre. 2017. 'The Cultivation of Entangled Listening: An Ensemble of More-Than-Human Participants'. In *Performance and Participation: Practices, Audiences, Politics*. Eds. Anna Harpin and Helen Nicholson. London: Palgrave Macmillan. 19–40.

Hosking, Rebecca. 2009. *A Farm for the Future*. BBC documentary. https://vimeo.com/136857929.

Hosking, Rebecca. 2021. '*A Farm for the Future*: How the Making of this Film Changed the Path of My Life'. https://www.rebeccahosking.co.uk/a-farm-for-the-future/.

Ingold, Tim. 2007. *Lines: A Brief History*. London and New York: Routledge.
Ingold, Tim. 2011a. *Being Alive: Essays on Movement, Knowledge and Description*. Abingdon, UK: Taylor & Francis Group.
Ingold, Tim. 2011b [2000]. *Perception of the Environment: Essays on Livelihood, Dwelling and Skill*. New York and London: Routledge.
Ingold, Tim. 2015. *The Life of Lines*. London and New York. Routledge.
Jones, Ffion, n.d.1. *Woollying the Boundaries*. Book 1: Fieldwork. Unpublished photographic book.
Jones, Ffion, n.d.2. *Woollying the Boundaries*. Book 2: Artistic Practice. Unpublished photographic book.
Jones, Ffion. 2012. *Ode to Perdurance/Awdl Amser*. [YouTube] https://www.youtube.com/watch?v=6FeJK_O-HWc.
Jones, Ffion. 2014. *Woollying the Boundaries: Perceptions of, and Interventions into, Upland Sheep Farming in Wales*. [PhD thesis] University of Aberwystwyth. https://pure.aber.ac.uk/portal/files/10625276/Jones_Ff.pdf.
Jones, Ffion. 2017. 'Agricultural Detritus and Artistic Practices: Reflecting on Animating Heritage and Reclaiming Place-Specific Narratives'. *Performance Research* 22.8: 115–126.
Lippard, Lucy R. 1997. *The Lure of the Local: Senses of Place in a Multicentered Society*. New York: The New Press.
Lorimer, Hayden. 2003. 'Telling Small Stories: Spaces of Knowledge and the Practice of Geography'. *Transactions of the Institute of British Geographers* 28.2: 197–217.
Lorimer, Hayden. 2005. 'Cultural Geography: The Busy-Ness of Being "More-Than-Representational"'. *Progress in Human Geography* 29.1: 83–94.
Myers, Misha. 2011. 'Vocal Landscaping: The Theatre of Sound in Audiowalks'. In *Theatre Noise: the Sound of Performance*. Eds. Lynne Kendrick and David Roesner. Newcastle-upon-Tyne: Cambridge Scholars Publishing. 70–81.
Nancy, Jean-Luc. 2007. *Listening*. Trans. Charlotte Mandell. New York: Fordham University Press.
Paolucci, Paul and Margaret Richardson. 2006. 'Sociology of Humor and a Critical Dramaturgy'. *Symbolic Interaction* 29.3: 331–348.
Pearson, Mike. n.d. *Carrlands Project*. http://www.carrlands.org.uk/images/carrlands.pdf.
Pearson, Mike. 2006. *'In Comes I': Performance, Memory and Landscape*. Exeter: University of Exeter Press.
Pearson, Mike. 2007. *Carrlands*. http://www.carrlands.org.uk/.
Pearson, Mike. 2010. *Site-Specific Performance*. Basingstoke: Palgrave Macmillan.
Pearson, Mike. 2011. 'Deserted Places, Remote Voices: Performing Landscape'. In *Envisioning Landscapes, Making Worlds: Geography and the Humanities*. Eds. Stephen Daniels, Dydia DeLyser, J. Nicolas Entrikin and Douglas Richardson. London and New York: Routledge. 280–286.

Pearson, Mike and Michael Shanks. 2001. *Theatre/Archaeology*. London: Routledge.

Pitches, Jonathan. 2020. *Performing Mountains*. London: Palgrave Macmillan.

Rancière, Jacques. 2009. *The Emancipated Spectator*. Trans. Gregory Elliott. London: Verso.

Rose, Mitch. 2006. 'Gathering "Dreams of Presence": A Project for the Cultural Landscape'. *Environment and Planning D: Society and Space* 24: 537–554.

Spirn, Anne Whiston. 1998. *The Language of Landscape*. New Haven and London. Yale University Press.

Tilley, Christopher. 2012. 'Walking the Past in the Present'. In *Landscapes Beyond Land: Routes, Aesthetics, Narratives*. Eds. Arnar Árnason, Nicolas Ellison, Jo Vergunst and Andrew Whitehouse. USA: Berghahn Books. 15–32.

Wilson, Louise Ann. 2016. *Emplacing, Re-Imagining and Transforming 'Missing' Life-Events: A Feminine Sublime Approach to the Creation of Socially Engaged Scenography in Site-Specific Walking Performance in Rural Landscapes*. [PhD Thesis] Lancaster University. https://eprints.lancs.ac.uk/id/eprint/85071/1/2017wilsonphdinternal.pdf.

Wilson, Louise Ann. *The Gathering/Yr Helfa*. https://louiseannwilson.com/work/the-gathering.

Wilson, Louise Ann. n.d. *The Gathering/Yr Helfa* and National Theatre of Wales. https://www.nationaltheatrewales.org/ntw_shows/gathering-yr-helfa/#gallery.

Wylie, John. 2002. 'An Essay on Ascending Glastonbury Tor'. *Geoforum* 33: 441–454.

Wylie, John. 2005. 'A Single Day's Walk: Narrating Self and Landscape on the South West Coast Path'. *Transactions of the Institute of British Geographers* 30.2: 234–247.

CHAPTER 4

Cultivating Dialogues Beyond the Farm

Interventions into Farmscapes of Global Food Chains in Agnes Denes' *Wheatfield---A Confrontation*, Nessie Reid's *The Milking Parlour* and Shelley Sacks' *Exchange Values*

The last two chapters explored the storytelling of a range of farmscapes through the farmers' words and agricultural practices and through the language of the farmscapes found in the traces visible on the land and asked how the artists used these performances to inform and engage the public as first steps in constructing a food democracy. This chapter looks at ways in which artists link storied farmscapes to global food chains and bring the public into the conversation through socially engaged art forms where the art is 'the process of communicative exchange rather than a physical object' (Kester 2004: 90). The shift to artwork as collaboration, participation and conversation, variously called the 'social turn', the 'relational turn' or the 'performative turn',[1] can be found in visual and performing art forms since the 1980s. Art historian Claire Bishop explains that in this changed relationship between artist, artwork and spectator,

[1] See, for example, Lacy 1995, Bourriaud 2002, Kester 2004 and 2011, Fischer-Lichte 2008, Bishop 2004, 2009 and 2012, Jackson 2011, Harpin and Nicholson 2017.

© The Author(s), under exclusive license to Springer Nature Switzerland AG 2021

S. C. Haedicke, *Performing Farmscapes*, Performing Landscapes,
https://doi.org/10.1007/978-3-030-82434-1_4

the artist becomes facilitator and collaborator of situations rather than sole creator of art objects; the artwork shifts from a portable commodity to a durational project relying on conversation; and the spectator takes on the role of participant-collaborator (Bishop 2012: 2). Nicolas Bourriaud, curator and art critic, proposes the term 'relational aesthetics' to describe these artworks in which social interaction becomes the art. As the exchanges are between artists as facilitators and spectators as co-creators, these artists, in effect, empower audiences to use the art to reflect on their particular social contexts and contribute to improving them. Bourriaud argues that these artists seek to collaborate with participants to alter small pockets of society, perhaps even just temporarily. 'The role of artworks is no longer to form imaginary and utopian realities, but to actually be ways of living and models of action within the existing real, whatever the scale chosen by the artist (Bourriaud 2002: 13). Bishop cautions against what she sees as the inherent assumptions of art's social turn: that socio-political engagement is synonymous with good art and that it necessarily forges a collaboration between artists and non-artists that is democratic and politically responsible.[2] Expressing her apprehensions about the value of sharing autonomy over the creation of the artwork with participants, she insists that participatory art be critically analysed through a lens of aesthetics, not simply ethics and efficacy. For her, 'the success of their work is not dependent upon authorial suppression, but upon the careful deployment of collaboration to produce a multi-layered event that resonates across many registers. As such, they think the aesthetic and political *together*, rather than subordinating both within the exemplary ethical gesture' (Bishop 2009: 250).[3] For art historian Grant Kester, the social turn in art with its shift in focus to social interaction and collaborative practices is not a radical break from the past. He locates its beginnings in the nineteenth century as artists shifted from idealising the status quo of political power dynamics with flattering images of the political elite to criticising and satirising them. He argues that this shift placed the artist in the position of provocateur and art in the role of opposing the dominant culture. That oppositional stance, in turn, alters

[2] Bishop 2009. This article was originally published as 'Antagonism and Relational Aesthetics' in *October* in 2004 and revised and published in *Artforum* in 2006. Bishop revised it to comprise the first chapter in *Artificial Hells* (2012) adding an excellent overview of key ideas and issues around this participatory art.

[3] Kester 2011 offers a critique of Bishop's claims, pp. 31–33 and 59–63.

meaning-making to prioritise discomfort or rupture as the status quo is challenged. That discomfort, Kester claims, causes a more active engagement on the part of audiences with the art—what he calls a 'decoding or somatic disruption' (Kester 2011: 36). This need to interpret the art established the foundation for what became increased audience participation in the artwork (Kester 2011: 29–36). For Kester, contextualising the shift to the social turn helps validate the aesthetic and political credentials of dialogic practices of collaborative projects. This dialogic aesthetic highlights communicative exchange and locates it in the social world where those involved engage with real-life issues:

> The effect of collaborative art practice is to frame this exchange (spatially, institutionally, procedurally), setting it sufficiently apart from quotidian social interaction to encourage a degree of self-reflection, and calling attention to the exchange itself as creative praxis. A particular experience of openness is encouraged as participants are implicated in an exchange that is not wholly subsumable to conventional, pragmatic demands, but is consciously marked as a form of artistic practice. In fact, it is in part the lack of categorical fixity around art that makes this openness possible. The distancing from the protocols and assumptions of normative social exchange created by aesthetic framing reduces our dependence on default behaviours, expectations, and modes of being, encouraging a more performative and experimental attitude toward the work of identity. (Kester 2011: 28)

A dialogic aesthetic is particularly helpful in articulating artistic processes of knowledge exchange and critical reflection as it clarifies the interactive practices that can lead to actual shifts on the part of participants from food *consumer* to food *citizen*, especially relevant to the case studies in this chapter. Each of these dialogic projects enables the participants to encounter the farmscape beyond the individual farm as each operates in social, political, economic and ecological landscapes of the global agri-food system. Through engaged dialogue and embodied activities, artists and participants in these case studies participate in collective artistic processes that investigate wheat, dairy and banana food chains. The aim of each of this chapter's projects is to provoke dialogue that provides an opportunity for sharing and debating ideas and practices among farming communities, agricultural experts, artists and the general public. The projects thus provide varied roadmaps to achieving food democracy as they develop food citizens who are knowledgeable about the agri-food system, pro-active in improving it, and concerned about the public good over individual gain (Fig. 4.1).

A pioneering example of this shift to dialogic art was Agnes Denes' public art intervention, *Wheatfield—A Confrontation*. Denes startled New York City residents in May 1982 with an agricultural landscape—a two-acre field of wheat—planted in the Battery Park landfill in lower Manhattan, two blocks from Wall Street and the still-standing World Trade Towers. After clearing the site of rocks and garbage and bringing in two hundred truckloads of soil, Denes and participants in the project dug two hundred and eighty-five furrows by hand and then sowed the seeds, again by hand, as curious residents came to watch. The field was maintained for four months, and the crop, harvested on August 16, yielded over 1000 pounds of healthy, golden wheat. *Wheatfield—A Confrontation* occasioned an unexpected encounter with farmlands that were *out of place* in the surrounding urban landscape and that challenged capitalist measures of the value of this piece of New York City real estate. In so doing, the urban farmscape cultivated an aesthetic-political dialogue with New York City residents as it exposed the social organisation and ideological operation of the city. Denes wrote of the project:

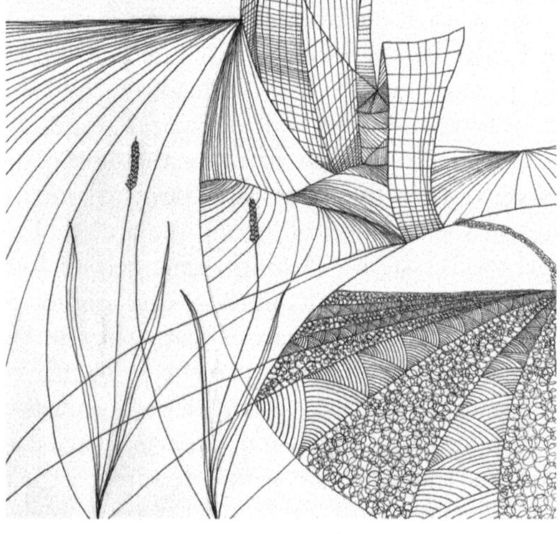

Fig. 4.1 *Horizon*. Pen and ink drawing by Kristen Oshyn

> Manhattan is the richest, most professional, and most congested … island in the world. To attempt to plant, sustain and harvest two acres of wheat here, wasting valuable real estate and obstructing the 'machinery' by going against the system, was an effrontery that made it a powerful paradox. It was insane. It was impossible. But it did call people's attention to having to rethink their priorities and realize that unless human values were reassessed, the quality of life, even life itself, was in danger... *Wheatfield* was a symbol, a universal concept (Denes n.d., in section entitled 'Writings').

The *art* of the project was more in the discussions it invigorated than in the object it constructed (the wheat field) as it became an aesthetic resistance to the way the city functioned and a 'call to care' (Rose 2006: 542) for the environment. The former landfill became the source of debate on urban land use determined by capitalist forces, but here the debate was guided by aesthetics and agriculture rather than profit-oriented city planning. As *Wheatfield—A Confrontation* highlighted urban power structures and policies that often remain hidden, it raised provocative questions about planting and harvesting a field of wheat on land worth $4.5 billion and then giving the wheat away for free.[4] It challenged the capitalist system on which the city thrived, and encouraged the public to counteract misplaced priorities by undermining the thrust of capitalist emphasis on profit and to draw attention to the power of prioritising the common good. In Denes' *Wheatfield—A Confrontation*, the grain, once harvested, travelled to twenty-eight cities around the world in an exhibition called 'The International Art Show for the End of World Hunger' (1987–1990), and then the seeds were given to people who planted them all around the globe. The project created a naturalcultural farmscape by asking whether prioritising individual/corporate profit was sustainable on

[4] Sarah Kavage's *Industrial Harvest* (2010–2011) followed a similar modus operandi to *Wheatfield—A Confrontation* of undermining capitalist profit goals by distributing twenty tons of flour for free. She exposed the impact of the US commodities market on the lives and eating habits of ordinary people by revealing how an abstract 'wheat futures' contract connected to real wheat, real food and the lives of ordinary people. Kavage used a futures contract that she bought on the Chicago Board of Trade to lock in a price of the wheat, bought 1000 bushels of real wheat that she had milled into flour and then gave it away to Chicago residents, food banks, homeless shelters, soup kitchens and schools as a way to de-commodify the wheat and transform it from a generic unit of investment back to real nourishment and to provoke dialogue about how the Board of Trade and the wheat commodity system was connected to what people eat and how much they pay for it. See Kavage n.d.a and n.d.b.

a fragile planet, and it challenged the push towards urban construction over the creation of green spaces.

The location of the working farmscape in the city in full view of the residents pointed out that detaching urban dwellers from the sources of their food led to ignorance of and indifference towards ethical food chains as Denes' temporary public art intervention wove embodied farmscape stories into the already existing urban narratives to disrupt the status quo. Cultural historian Rosalyn Deutsche looks at public art as a practice that 'participates in the production of meanings, uses and forms for the city' (Deutsche 1996: 56), and she argues that '[a]rt that is "public" participates in, or creates, a political space and is itself a space where we assume political identities' (Deutsche 1996: 289). Art historian Patricia Phillips also dismisses the notion that public art is just art in public spaces and insists that '[i]t is public because of the kinds of questions it chooses to ask or address' (Phillips 1992: 298).[5] Public art can play a significant role in the construction of the meanings and uses of public spaces by making them sites of debate, and without the exchange of ideas and controversies, public art no longer fulfils its purpose. Phillips argues for an ephemerality in art in public spaces to reflect these dynamic social practices of public life:

> The temporary in public art is not about an absence of commitment or involvement, but about an intensification and enrichment of the conception of the public. ...A conceptualization of the idea of time in public art is a prerequisite for a public life that enables inspired change. (Phillips 1992: 304)

Denes has recreated *Wheatfield* in the East End of London in 2009 and in downtown Milan in 2015. In each version of the performance intervention, the farmscape that emerges goes beyond the farm to become a communicative exchange on land use, food production, agricultural sustainability and food security.

Like *Wheatfield—A Confrontation*, Nessie Reid's *The Milking Parlour*[6] inserted a thought-provoking public art event in the form of

[5] See also Phillips 2003.

[6] See Reid 2015, 2016a, 2016b, and 2016c.

an iconic farmscape into a city centre[7] transforming an urban landscape into an urban–rural hybrid with political overtones. In both these projects, the significance of the local is complicated as the *local* of the farm is transported to the *local* of an urban setting so that the *local* becomes a juxtaposition of rural and urban, of farm and city public space. This, in turn, draws attention to socio-political challenges of city/country dynamics. In April 2016, British performance artist and political ecologist Nessie Reid set up *The Milking Parlour* in Bristol city centre where she lived for five days in a pop-up milking parlour with two Guernsey cows, Meadowsweet and Alisa from Graylands Farm in Dundry near Bristol, UK. Passers-by in Bristol's busy harbourside unexpectedly came across a fenced area covered with hay. Inside the fence was a shed for the two cows and Reid who stayed with them day and night, milked them, fed them and mucked out the shed every day, but she also engaged in conversations with the general public about the difficult realities of milk production and its impact on climate change. On the fence surrounding *The Milking Parlour* site were several information panels with facts about the dairy industry and how the low cost of milk makes it difficult for dairy farmers to survive. Every morning, she placed two blank blackboards and chalk outside the fenced area: one for the public to write questions to ask the farmers, the other for farmers to ask questions to the public. These questions became the starting points for the daily afternoon discussions as dairy farmers, vegans, academics, agricultural experts and the public sat in a circle on hay bales and talked with each other, bridging the gap between farming and non-farming communities.

The Milking Parlour was supported by Cape Farewell, an international not-for-profit programme based in the UK, that funds artists who address issues of climate change through a cultural lens. Like Denes' *Wheatfield—A Confrontation*, Reid's durational street intervention asked the public to consider what sustainable farming would look like. *The Milking Parlour* did not seek to offer a romanticised picture of cows and farming or to give families a quasi-farm outing, Reid insisted, but rather to highlight the hard work—the working farmscape—in caring for the cows and producing such an ordinary item on supermarket shelves as milk. Reid

[7] Other more whimsical outdoor performances also bring farmscapes to urban public spaces to encourage dialogue with the public about farming and food production, notably Opéra Pagaï's *Entreprise de Détournement*, Le Phun's *The Seedlings' Revenge*, and Friches Théâtre Urbain's *Aroma-Home*. See Haedicke, 2012, 2016, and 2018.

sought 'to provide an inquiry into the values we hold, up-hold and could do without relating to the food and drink we buy' (Reid 2015: n.p.) and to draw attention to the environmental implications of industrial agriculture for our planet. George Ferguson, Mayor of Bristol at the time of the art installation, explained, 'Nessie has brought these cows as an artwork, but also as a provocation, I think, to make us think about where our food comes from. This connection between country and city is vitally important. Too many of our children and many of the population think that food comes off supermarket shelves' (Reid 2016b). Always a milk drinker, Reid admitted that she had not really thought about its journey from the dairy farm to the supermarket shelf until she attended a Sustainable Food Trust conference on the true cost of food and farming in 2013. There she learned about the environmental, animal welfare and human health costs of cheap food as speakers claimed that it is more profitable to farm unsustainably than sustainably. She began to realise that the quest for cheap food had distorted the public's understanding of the real value of food, blatantly evident in the crisis in the dairy industry with the precipitous fall in milk prices making milk cheaper than bottled water. In *The Milking Parlour Film*, she explained:

> Now cheaper than bottled water, milk is the perfect lens to explore some of the broader personal and societal values that we hold toward our food system. Agriculture is one of the largest contributors to climate change and biodiversity loss, but not all forms have to be destructive. I want to know how we feed the planet with its burgeoning population without it costing the earth and the lives of our generations to come. … [Milk] is a useful medium to choose, but what I really want [*The Milking Parlour*] to be about is for us to explore our connection and our relationship with food. … What we eat, what we choose to eat, what we choose to buy in the supermarket is almost a political choice … We're making ethical choices and choices that serve the people and the planet. That is a massive contribution to society. (Reid 2016b)

While many people welcomed *The Milking Parlour*, some animal rights activists called it a 'publicity stunt' that ignored the realities of milk production and would cause undue stress to the cows. Reid welcomed opposing opinions both in onsite and online conversations. The information panels, blackboards, informal conversations with the public and the daily discussions about the dairy industry certainly provoked dialogue as *The Milking Parlour* challenged participants to think about dairy food chains and the tangible costs that the low cost of milk has on farmers as well as the impact of dairy farming on the environment, and it asked the public to be part of the solution by being milk *citizens* rather than just milk *consumers*. Reid transformed a public urban space into a rural dairy farm, albeit temporarily, to raise awareness about the current state of milk production in the UK, 'to spotlight the disconnect between the cities we live in, the food we eat and the milk we consume on a daily basis' (Reid 2016a), and to create a discursive space that encouraged the public to have a say in the construction of their food systems.

Like Denes, Reid created an ephemeral public art intervention to provoke discussion and reflection among urban residents who may have very little knowledge of farming and food production by inserting agricultural practices into an urban environment. Unlike Denes, however, Reid made conversation an integral part of the performative event as she was always on site to talk to curious passers-by. She claimed that this liveness, both from the conversations and her interactions with the cows, was crucial to engaging a seemingly uninterested public. Reid explained that when we participate in a live event:

> when we have an embodied physical experience, we relate to it more strongly…. Images of slaughter houses or fibreglass cows wouldn't have had the same impact. People could smell the cows, see them peeing and pooing, and see me milking them, as well as get a sense of the connection I had with the animals. I would rather give that direct experience to a few thousand people than to a million who couldn't connect to it. (Braham 2016, n.p.)[8]

The Milking Parlour has been performed in multiple locations for the last several years, sometimes without Meadowsweet and Alisa, and it won the P.E.A. (People, Environment, Achievement) Award in 2016.

Like Nessie Reid, Shelley Sacks developed a project that engaged the public in the wider debate about the landscape of the global agri-food system and trade agreements that prioritised low cost food over economically and environmentally sustainable practices. For both these artists, time and space for dialogue were integral parts of the artwork and highlighted the significant role of conversation in the formation of food citizens. Although Sacks moved her performative exhibition into a

[8] See also Ruscoe 2016, n.p.

museum or conference setting, she, like Reid, relied on a sensory, or 'live', approach to engage her audience as she stimulated the spectators' senses of hearing and smell. In Sacks' *Exchange Values*,[9] that has toured the world since 1996, audiences are confronted with the smell of thousands of dried banana skins and reproached by the voices of banana farmers on St. Lucia in the Windward Islands who supply the UK with the majority of their bananas. Ian Cook who saw the event in 2000 explained:

> They're in your face. Banana skins. Dried. Cured. Blackened. Flattened. Sewn together in a panel. Stretched. Taut. On a frame. Right in your face. And it smells. It's rich. Gorgeous. You can't move too far away. Get too distanced. If you want to keep the headphones on. The ones attached to the little metal box below the panel and the frame. The one with the number on. (Cook 2000: 338)

In a large exhibition room, twenty panels of stitched together dried banana skins were carefully hung on the walls. Under each panel was a metal box with a stencilled number and a set of headphones. Each number was a banana grower's number (for example, M330534), so each box metaphorically contained a St. Lucia farmer who had recorded a short narrative on his or her banana farming experiences. Visitors to the exhibit moved from one local, working, naturalcultural farmscape story to another, listening to the farmer's voice and smelling the strong odour of the dried banana skins from his or her specific farm. For the first decade of showings, a large carpet of 10,000 unnumbered banana skins—the countless nameless and voiceless farmers—filled the centre of the room. In 2007, Sacks replaced the carpet with a round table, five meters in diameter, with the 10,000 banana skins now filling the centre of the table. Although discussion had always been a part of the exhibit, this switch from carpet to table and benches, as well as a shift in the title of the exhibit from *Exchange Values: Images of Invisible Lives* to *Exchange Values: On the Table*, confirmed the increased emphasis Sacks put on dialogue to create active food citizens. The addition of the table and its sense of an ongoing conference reminded the participants that they could not be passive visitors to the global food system able to participate at will. Rather, they are

[9] See Sacks 2016, *Exchange Values* website, for descriptions and photographs. All of the primary source material on *Exchange Values* is on the website in various sections, so I have cited the specific urls. See also Heim, 2005.

inescapably entwined in its complex network as both banana consumers and as banana citizens who have a responsibility to be informed, to participate in the banana food chain through their demands for ethically sourced bananas and their complaints that keeping the price of bananas so low was unfair to banana growers, and to fight for justice for the farmers in terms of fair wages, reasonable working conditions and environmentally sustainable agricultural practices.

The disembodied voices of the farmers, confronting the audience through the headphones, vent their frustrations with their difficult working conditions, low pay, the power of the multinational corporations to set prices and the general public who are pushing banana prices so low that it is hard for the farmers to survive. 'We down here have quite a bit of work to do to produce a bunch of bananas and what are we getting in return is far too little. ...More work and less money!' (Sacks, A320129[10]) says one farmer into the listener's ears. Another complains that 'we sometimes have to produce bananas and get payment below the cost of production' (Sacks, R610145). A single mother with ten children angrily says 'Farming is not something easy. It is very difficult.... We want more money. They ask for quality; we providing quality but at the same time we want a payment for the quality....I want to make the people of Europe to know how difficult we are taking it [sic.]' (Sacks, A320129). The St. Lucia government recommended that farmers diversify into tourism that brings 'top dollar' to supplement their income, but the farmers resented that suggestion often expressing that they did not want to be slaves again cleaning people's shoes. By and large, the farmers were 'angry and aware that the fluctuating prices and the fates of their families have little to do with bananas or how hard they work. Some growers blamed the government, others the "growers associations". Still others talked about multinationals playing "a money game"'.[11] One farmer asked: 'Where is the money going?... Give the farmers a chance'

[10] All the quotations with parenthetical documentation like this one indicate that the quotation is from one of Sacks' recorded interviews with the banana farmers that could be heard in the *Exchange Values* exhibit. The interviewee is identified by the grower identification number that appears on the box of bananas shipped abroad. A recording of each of the twenty interviews is available on the *Exchange Values* website.

[11] Sacks, 'Collecting the numbered skins', http://exchange-values.org/uncategorized/story-of-the-project-2/.

(Sacks, E490347). As the farmers spoke, they seemed to acquire an ethereal presence through the banana skins. For the farmers, the banana skins and their own life stories are synonymous: 'You can see from this "skin" that a banana is not an easy thing. It has our lives in it. Our culture. Our struggle. Maybe people will see that even if we love bananas, we can't go on like this anymore. Something has to change'.[12]

The first stage in Sacks' creative process of this ongoing project[13] was *Reading the World Economy in the Banana Skins*, more of an experimental practical intervention research activity than a fully developed performative project. Sacks admitted:

> I began drying banana skins around 1970, not for any specific purpose, but because I found it hard to throw them away. I would stand with a skin in my hand, wondering where it had come from, who had grown it, what the life of this person was like. Each skin still had so much life in it; it seemed a pity to throw it away. So I stretched strings across the wall of my room, where the skins could hang to dry. As they dried, blackening, twisting, stiffening, they would speak through their silent forms.[14]

When Sacks moved to Germany in the early 1970s to study with Joseph Beuys, she brought her dried banana skins in a wooden trunk. Over the next few years, she intervened into the everyday activities of passers-by on the city streets by placing the skins on a mat on the pavement and, following in the footsteps of her grandmother who read tea leaves, she would 'read the world economy in the banana skins'[15] to initiate conversations with those who stopped to listen and respond. Thus began a project that continued to develop in depth and scope to today, fifty years later.

[12] Ibid.

[13] See Sacks, 'Evolution from 1970', http://exchange-values.org/story-of-the-project/history-of-the-project/; 'Invisible Lives 1996–2006', http://exchange-values.org/story-of-the-project/images-of-invisible-lives-1996/; 'Collecting the numbered skins', http://exchange-values.org/uncategorized/story-of-the-project-2/ and 'On the Table 2007 Onwards', http://exchange-values.org/story-of-the-project/on-the-table-2007/.

[14] Sacks, 'Invisible Lives 1996–2006', http://exchange-values.org/story-of-the-project/images-of-invisible-lives-1996/.

[15] Sacks, 'Evolution from 1970', http://exchange-values.org/story-of-the-project/history-of-the-project/.

The research and development for what became the exhibit, *Exchange Values: Images of Invisible Lives*, started over twenty years later in the early 1990s as the GATT free trade agreement and its impact on banana growers in the Windward Islands was frequently the news. Buying bananas one day, she saw a 'grower identification number' on one of the boxes and wondered whether she could find the actual growers of individual boxes of bananas. She imagined creating a piece in which UK banana consumers could begin to understand where their bananas came from and who, in particular, planted and harvested them. Getting to know the farmer and the challenges of banana farming, Sacks believed, would encourage members of the public to see the role played by multinational competition in driving prices down often below a living wage and that, in turn, would encourage them to take some responsibility for an ethical food chain and shift their role from consumer to citizen in a food democracy (although Sacks does not use these words). In June, 1996, after three years of research and discussions with government officials, industry executives and representatives of the Windward Island Banana Growers Association, Sacks was ready to start collecting skins from numbered boxes. On her website, she details the banana skin collection process as she and four collaborators handed out three thousand bananas for free from twenty growers along with a printout of the grower's number to passers-by in Nottingham's city centre. She just asked that they eat the banana right then and return the skin to her. The process of drying the skins occurred immediately so that the skins would not develop mould. The website offers details on how the skins were dried and stitched into the panels that would eventually hang on the walls of the exhibit. Preparing the banana skins was followed by finding the twenty growers in St. Lucia whose banana skins she had dried, and in July, 1996, she went to St. Lucia to gather their stories. Influenced by Joseph Beuys' ideas on *social sculpture*,[16] Sacks sought to develop a 'large scale "social sculpture" that lifted the creative process out of the specialist world of art and placed it into a realm in which we all have the possibility to become artistic "social sculptors", engaged in shaping our world'.[17] The performative exhibition with its twenty panels of dried banana skins

[16] See Jordan 2013 and Rojas 2010.

[17] Sacks, 'Collecting the numbered skins', http://exchange-values.org/uncategorized/story-of-the-project-2/. See also Heim who says that this performance exhibition is 'a "model process" of social sculpture' (2005: 204). See also Beuys 2006 and Rojas 2010.

and the stories the farmers chose to tell in their own words about growing and harvesting those bananas became the foundation of her social sculpture as she sought to re-sculpt the social world of the banana industry by enhancing consumer awareness and enabling knowledge exchange in the public dialogue surrounding the exhibit. Each of the eleven exhibitions of *Exchange Values* since 1996 took place as part of a conference or a larger art exhibition that looked at some aspect of global trade or sustainable development, so lectures and discussions were woven into the exhibit that was, in turn, placed in a larger global context. For example, the exhibition at the Johannesburg Art Gallery, South Africa, lasting from August to December 2002, became part of the 2002 World Summit on Sustainable Development.

The conversations among the public who came to hear the farmer's life stories and smell the banana skin panels were certainly most visible in the colloquia connected to the exhibition, but behind them were the hundreds of conversations with supermarket managers, government officials, policy makers, artists, academics and, of course, the farmers. These conversations did not take place just to enable the completion of an artwork, but rather were dialogues that tried to discover causes and seek accountability for the dire plight of the banana farmers. The recordings enabled the farmers to 'speak' to the eaters of their bananas and to describe the impact that the demand for cheap and perfectly shaped bananas has had on their farms in terms of increased hours needed to make the same income as when bananas were more expensive, the harmful chemicals needed to produce those bananas, and the devastating effect of free trade agreements. And the discussions among activists, artists, farmers' representatives, educators and consumers surrounding the event addressed political, social, economic and ecological issues around agriculture and the public's relationship to food and explored ways to change the banana food chain. For Sacks, these conversations:

> confirmed that the aesthetic—as the opposite of numbness, of anaesthetic—is closely linked to our ability to respond. In this space—beyond the linear, the literal, the discursive—where the creative imagination weaves and moves, we are moved inwardly and become internally active. Responsibility then has less to do with moral duty, and becomes rather an ability to respond. This kind of response-ability enables us to care at a distance,

and to develop the will to act in accord with what we now understand and see.[18]

Her use of the word 'response-ability' echoes Donna Haraway's use of the word (discussed in Chapter 1) and brings with it the same compulsion to listen and nurture others. Haraway claimed that response-ability means using one's ability to respond to the needs of others: 'a collective knowing and doing, an ecology of practices. Whether we asked for it or not, the pattern is in our hands. The answer to the trust of the held-out-hand: think we must' (Haraway : 34). Sacks does not claim her artwork necessarily leads to direct action, but the stories it told could cause emotive shifts: 'For example, the experience of absence is so tangible—of a producer whose "skin" is stretched before us, whose voice is inside us—that it stirs one imaginatively, provoking an inward movement that we carry outwards into the world'.[19] *Exchange Values* certainly increases consumer knowledge of the story of banana production through the recorded voices of the farmers and the sight and smell of the banana skins. With the addition of the round table in 2007, the importance of dialogue was reinforced, creating what Beuys called a 'permanent conference'. While the website for the project claims efficacy for *Exchange Values* in the dialogues it promotes, the banana industry has remained largely unchanged. It is impossible to assess the impact on the many audience members at the performative events of Denes, Reid or Sacks, but the engagement of the public in the agri-food system has certainly increased since the 1980s when Denes planted her first field of wheat. Projects like these contribute to the formation of food democracy by creating spaces where conversation and knowledge exchange among artists, the general public and agricultural experts can inform all involved and where audiences can rehearse citizenship by using art to imagine a more sustainable agricultural future. This problem-posing dialogic approach is developed further in the next case study, *A Field of Wheat*, where participants are given the responsibility to become food citizens through dialogue and decision-making on the farm.

[18] Sacks, http://exchange-values.org/wordpress/wp-content/uploads/2016/07/shelleysacks1.pdf.

[19] Sacks, http://exchange-values.org/wordpress/wp-content/uploads/2016/07/shelleysacks1.pdf.

Cultivating Food Citizenship around Wheat: Anne-Marie Culhane and Ruth Levene's *A Field of Wheat: Arts and Agricultural Project*

A Field of Wheat: Arts and Agricultural Project (2015–2016) was a dialogic, collaborative art-based project involving a farmer in Lincolnshire (Peter Lundgren), two artists (Anne-Marie Culhane and Ruth Levene), the Collective ('forty-two members of the general public, the food industry, the farming community, artists and researchers'[20]), and the field of wheat, called Middle Field, on Lundgren's farm in Branston Booths, Lincolnshire, England.[21] Anyone could join the Collective by paying a one-time membership fee. Once the wheat was sold, Collective members could receive a payment that was dependent on the wheat's sale price, so members could lose money, break even or make a profit, just like the farmer. Being a member of the Collective, of which I was one, gave us the opportunity to encounter the actual wheat field on Lundgren's farm, engage in the embodied art practices and farming tasks, encounter the farmscape beyond the farm through its online presence on an interactive website and join a conversation around wheat farming in the larger agricultural context. Being a member of the Collective also meant that we agreed to be active participants in the project through knowledge exchange and decision-making. The members of the Collective were simultaneously spectators of the performative events in the art-agriculture project, participants in the project's on site and online activities and an engaged community working on a participatory action research project around sustainable wheat farming. In *A Field of Wheat*, dialogue in multiple forms was at the heart of the project. As Levene explained, the project was about voices—listening and talking to the farmer, sharing knowledge, establishing dialogues between the farmer and the members of the public in the Collective, hearing different voices in the story and 'above all, listening to the land' (Personal interview, 16 March 2019). And for Culhane, the project created 'a parallel between the growing of the wheat in the field and the journey of the Collective' (Personal interview, 1 March 2019).

[20] The artists requested that the description of the participants be consistent with how they defined them previously. Email exchange, 27 May 2020.

[21] See *A Field of Wheat: Arts and Agriculture Project*, n.d.; Levene, 'A Field of Wheat with Anne-Marie Culhane' n.d.; and Freemantle, n.d.

The performative project of *A Field of Wheat* followed the wheat cycle from drilling the wheat in October 2015 to its harvest in September 2016. Levene and Culhane wrote: 'Our research focused in Lincolnshire because we wanted to go to the heart of UK "prairie country", the bread-basket of the UK'.[22] Since 95% of UK wheat is grown with fertilisers and pesticides, the artists, alongside the Collective members, decided to focus on that conventional wheat farming method to begin to understand the nation's dominant form of wheat production. But the project was not a one-way education process with the farmer imparting his wheat-growing expertise. Lundgren often said he learned so much about public attitudes towards conventional farming and passions about reducing the use of chemical pesticides and fertilisers in conversations with the members of the Collective. In addition, he received valuable information about differences in the soil in various areas of his field from Dr Tom Powell who tested it for nitrogen and carbon content. The dialogical aesthetic of *A Field of Wheat* 'open[ed] up a new space for informed discussion, reflection and sharing of different ways of thinking about wheat farming, food systems and Collective ownership, all within the larger context of resources depletion, climate change, global markets and technological innovation'[23] in order to 'point out the conflicts, contradictions, complexity, limitations and functionality or dysfunctionality of the current food commodity system'.[24] As active stakeholders, the Collective members and the artists participated in discussions about specific farm practices on Middle Field that placed Lundgren's field in a larger historical and current-day agricultural context and in four face-to-face onsite events, particularly the Field Trip to London on 4 March 2016 designed to help the Collective understand wheat trading since 1600 (discussed later in the chapter). Participation in the discussions continued via an extensive online platform. In addition, the Collective and the farmer together made real-life decisions that had an impact on the growing of the wheat. This interdisciplinarity of ideas and practices around farming, social and ecological justice, sustainability, performative and participatory art and

[22] 'Artists' Pages: Getting There', http://fieldofwheat.co.uk/artists-pages/getting-here/. The website is detailed and extensive, and much of the information on the project resides there, so I will include the specific urls to the various sections of the website in the footnotes.

[23] 'About the Project', http://fieldofwheat.co.uk/about-the-project/.

[24] 'The Exchange: Decision No. 1', http://fieldofwheat.co.uk/decisions/decisionno1/.

knowledge exchange were built into the project's complex structure that created an interactive durational performative event lasting one year.

To make sense of all the varied storylines in *A Field of Wheat*, it is helpful to turn to Tim Ingold's discussion of lines. Ingold argues that 'lines are everywhere. As walking, talking, gesticulating creatures, human beings generate lines wherever they go' (Ingold 2007: 1). In fact, he asserts, 'the world consists, in the first place, not of things but of lines' (Ingold 2007: 5) and sociality or 'clinging together'—a prevalent characteristic of human and nonhuman worlds—is 'an entwining of lines. … Nothing can hold on unless it puts out a line, and unless that line can tangle with others' (Ingold 2015: 3). Ingold identifies one type of line[25] as a thread, 'a filament of some kind, which may be entangled with other threads or suspended between points in a three-dimensional space' (Ingold 2007: 41). Threads have surfaces and create their own presence. In *A Field of Wheat*, the *physical thread* was obvious, for example, in the roots, stalks, leaves and spikes of the wheat plant or in the spider webs suspended between plants. Illustrations of some of these threads appear in the Artists' Pages on the website in 'Beneath Your Feet', a series of drawings of the growing wheat's roots over the year by Culhane, and in 'Ears for Listening', a series of photographs of the wheat's spikes by Levene.[26] Both Ingold and *A Field of Wheat* also introduce more metaphorical notions of lines as narrative threads of text and conversation, of storylines and life-lines, and of lines of inquiry and knowledge exchange. To understand how these multiple lines entangle, Ingold introduces the metaphor of 'knots'. The notion of a local place, he argues, is the sum of crossing paths that humans and nonhumans make as they move around and through a place and so create a 'complex knot [that] does not contain life but is rather formed of the very lines along which life is lived. These lines are bound together *in* the knot, but they are not bound *by* it' (2007: 100). The paths or life-lines lead into and out of the knot and entangle with other lines on similar but varying paths. It is these knots that constitute the social life of local place: 'in a world of *life*—knotting is the fundamental principle of coherence. It is the way forms are held together and kept in place within what would otherwise

[25] Ingold identifies another type of line as a trace (discussed in Chapter 3).

[26] See http://fieldofwheat.co.uk/artists-pages/. See also 'Artists' Pages: The Straight Line, Pt 1', http://fieldofwheat.co.uk/artists-pages/the-straight-line/ and Artists' Pages: The Straight Line. Pt 2', http://fieldofwheat.co.uk/artists-pages/straight-line-pt2/.

be a formless and inchoate flux' (Ingold 2015: 14). Ingold reminds his readers that knots exist only after the act of knotting, of tying together two or more physical or metaphoric threads that must be:

> linear and flexible. They do not meet face-to-face, on the outside, but in the very interiority of the knot, and they are joined neither end-to-end nor side-by-side but in the middle. Knots are always in the midst of things, while their ends are on the loose, rooting for other lines to tangle with. (2015: 22)

The threads in these knots are constantly in process of knotting and unknotting, constantly expanding and unravelling. Ingold's words summarise the performative tying, loosening and tying again of the varied lines of inquiry, knowledge acquisition and participation in *A Field of Wheat*, to create a complex knot of the farmscape, the wheat food chain and the project (Fig. 4.2).

These storied threads, some tangible, some metaphoric, constituted the core of this art-agriculture project. A key thread of *A Field of Wheat*

Fig. 4.2 *The Inextricable*. Pen and ink drawing by Kristen Oshyn

was the agricultural storyline as the 'script' followed the seasonal cycle of a seemingly barren field as it transformed into a 'field of wheat' that was then harvested. It was a performance of an annual wheat-growing cycle and its necessary accompanying agricultural tasks and practices. But knotted to that thread were so many other threads of inquiry and stories. Some of the threads can be followed along a chronological path, but others are more thematic or practice-based. I will seek to identify the paths of some of these knotted threads, not to untangle them, but to understand how they are tangled in the innovative artistic practices in this project that modelled a possible path to food democracy through threads of dialogue, knowledge exchange, networks, decision-making, re-imagining organisational roles in the food system and working towards a common good.

For Levene and Culhane, lines were important early in the research process. Two of the Artists' Pages look at the significance of lines[27]:

> Navigating blindly through the huge expanse of mono-culture was something that resonated throughout. Who is creating the lines? Are they slow meandering ones that bend and curve? Or direct straight ones where we remove the things in our way? What marks and scars do they leave on the land? Can we see where we are heading?[28]

At The Collection Museum in Lincoln, Culhane and Levene found George F. Carlisle's 1897 painting, 'The Ploughing Match' that depicts a crowd intently observing a contest among farmers to show off their skills to plough the straightest line. That discovery led them to look into scientists' findings that humans and animals cannot walk in a straight line without visual markers to keep them on track. They tested the theory by blindfolding Culhane who then tried to walk a straight line in Lincolnshire wheat fields. Invariably, she ended up walking in circles.[29]

[27] 'Artists' Pages: The Straight Line, Pt 1', http://fieldofwheat.co.uk/artists-pages/the-straight-line/ and 'Artists' Pages: The Straight Line. Pt 2', http://fieldofwheat.co.uk/artists-pages/straight-line-pt2/.

[28] 'Artists' Pages: The Straight Line, Pt 1', http://fieldofwheat.co.uk/artists-pages/the-straight-line/.

[29] 'Artists' Pages: The Straight Line, Pt 1', http://fieldofwheat.co.uk/artists-pages/the-straight-line/. See also Gonzalez, 2013 and Krulwich, 2010.

Fig. 4.3 The Alternative Ploughing Match. *A Field of Wheat.* 10 September 2016. Photographer: Nathan Gibson

Levene developed this exercise into an activity in the final face-to-face harvest event—the Alternative Ploughing Match. Here, participants covered their heads with flour sacks and walked in what they thought was a straight line only to discover their paths were filled with curves and spirals[30] (Fig. 4.3).

These lines and threads were not only in the fields; they were also tied to the larger agricultural context as they explored causal lines explaining Lincolnshire's fertile soil, the impact of fertilisers and pesticides and global distribution of wheat. Culhane and Levene explain:

> We had a shared curiosity about what it would feel like to be in and move around these extensive flatlands, to think about the connections between these fields and the food that is on our tables, to explore the place of these vast fields in a resource depleted, low carbon world. These vast

[30] 'Artists' Pages: The Straight Line. Pt 2', http://fieldofwheat.co.uk/artists-pages/straight-line-pt2/ for a description, photographs and drawings of this event.

> 'invisible' landscapes are often devoid of human presence – we wanted to bring people to these spaces and make them visible. Our research built on our existing knowledge and deepened our understanding of the cultural history, stories and customs that surround wheat farming, in Lincolnshire and beyond.[31]

After two years of preliminary research, they were ready to get to know Middle Field. Levene focused on the ditches and drains in the field which led to a mapping exercise that, in turn, revealed a buried water pipe system and pointed to an historical narrative thread on the ditches leading back to the fen slodgers, the area's early inhabitants before the fens were drained, who navigated their water world by pole vaulting (Personal interview, 16 March 2019).[32] Culhane sought to engage with the field through embodied practices like 'field sensing', a slow walking process to get a somatic sense of the field's size and topography.[33] It represented a kind of mapping through the experience of physical movement across Middle Field. For example, she measured the length of the field using a Buddhist prostration, an action that included taking two steps and lying down and getting up over and over. She admitted it took two hours and twenty minutes and gave her a sense of euphoria, like a pilgrimage, that, she believed, opened her to an intimate understanding of the nuances of Middle Field (Personal interview, 1 March 2019).

For the artists, getting to know the field was not about creating a concept or an image of wheat farming; they did not want simply to document the threads they gathered in written and visual forms for others to experience at a distance. Rather, they used their lived experiences, participatory research and embodied practices to understand what specifically was 'going on' in Middle Field in the context of the larger agricultural issues and to create a performative participatory project that reworked those lived experiences with the Collective. It is a landscape, Culhane explained, that non-farmers rarely experience except at the end of the food chain as consumers and eaters, so the project sought to invite the members of the Collective to enter this monocultural farmscape literally

[31] 'Artists' Pages: Getting There', http://fieldofwheat.co.uk/artists-pages/getting-here/.

[32] See Ruth Levene's description of fen slodgers in the Artists Pages, http://fieldofwheat.co.uk/artists-pages/in-the-name-of-progress/.

[33] See http://fieldofwheat.co.uk/artists-pages/using-body-map-field/.

with our bodies to explore what it felt like, what sensations and experiences we could have in that place. Only by physically entering this wheat farming system could we begin to understand what it means to grow food in this way on this farm, to appreciate how challenging it can be for the farmer, and to sense the role of this field in the wheat futures markets in global food systems (Personal interview, 1 March 2019). The project sought to reveal the story of wheat as grown in Middle Field while still locating that specific local field in the context of agriculture's global impact on climate change and the challenges around food production and food security today, so it explored all three modalities of local, working and naturalcultural farmscapes introduced in Chapter 1.

The complex dramaturgical structure of *A Field of Wheat* superimposed a trajectory of varied face-to-face and online conversations onto the annual wheat cycle in the field. The project's four face-to-face events encouraged embodied and emplaced conversations that coincided with the agricultural cycle.[34] The one face-to-face event occurring off the farm located the actual and virtual farmscape in an historical and socio-political context of the global wheat food chain. The Collective gathered for a walking tour of central London on 4 March 2016 as members learned about the history of wheat trading in the UK from the early seventeenth century to today and followed the journey of the wheat once it left the farm by visiting key landmarks in that story.[35] The walk started at Bear Quay, near Tower Bridge, where the sloops filled with the harvested grain were unloaded. The spot became an impromptu market until the amount of trading outgrew this surprisingly small area paved with stones slanting down to the River Thames. We then visited what used to be the Corn Exchange at 55 Mark Lane, now an office complex, but a reading from 'The Sketcher in London', taken from the publication, *The Leisure Hour* (1856)[36] that detailed a day at the Exchange, imaginatively transported us back in time as it vividly described the grain that covered the floor of the Corn Exchange growing in depth over the day:

[34] For photographs, see http://fieldofwheat.co.uk/newsandevents/.

[35] For a detailed description of the day's events and the texts read at the various sites, see pdf outline of the day under Field Trip to London, 4th March 2016, http://fieldofwheat.co.uk/newsandevents/

[36] 'The Sketcher in London' excerpt was included in the day's schedule, http://fieldofwheat.co.uk/wp-content/uploads/2016/03/londonfieldtripforweb.pdf.

> Grain lies scattered about in heaps everywhere, and on everything; it literally covers the ground ankle-deep, and we, and the whole ever-moving crowd, are trampling and grinding it under foot every step we take. How that comes to pass we see plainly enough, because every man who puts his fingers into a sample bag and takes out a handful, after rubbing it and examining it in his palm, instead of returning it to the bag throws it on the floor.

Rather than being wasteful, this practice ensured that the wheat tested by the next potential buyer was not tainted by the moisture of previous hands. The grain from the floor was swept up at the end of the day to become animal feed. Lundgren then read from Thomas Hardy's *The Mayor of Casterbridge* to locate the farmer in the wheat trading story.[37] Even though we were standing in the spotless lobby of the office complex, the sounds, smells and sights of the Exchange were palpable. We then walked past Bank to Bread Street, London's early bread market dating from 1302, where we shared a loaf of sourdough bread baked by one of the Collective members. At the final stop before lunch, we listened to the personal experiences of Jonathan Cowens who had worked in the wheat trade as CEO of Home Grown Cereal Authority for thirty-five years. Each stop enabled us to imaginatively follow the journey of the wheat in this farmscape beyond the farm and reinforced the local–global links.

Three of the events took place on the farm. The first event on 6 November 2015 focused on the land and its soil[38] with talks on the history of the local fens and the evolution of the soil[39] as members of the Collective walked the farm's fields rarely open to the public. The walk ended with a chance to vault a ditch using a pole like the fen slodgers.[40]

[37] For the complete texts read at the Corn Exchange, use the links in the pdf outline of the day under Field Trip to London, 4th March 2016, http://fieldofwheat.co.uk/newsandevents/.

[38] For photographs and descriptions of each of these events, see http://fieldofwheat.co.uk/newsandevents/.

[39] 'These include the dark fenland peat, formed when the fens were flooded marshes, sticky clay soils deposited by ancient tidal rivers, and lighter silty soils that mark networks of slow moving creeks', http://fieldofwheat.co.uk/wp-content/uploads/2015/09/firstfarmeventblog.pdf.

[40] See 'Artists' Pages: Fen Slodgers', http://fieldofwheat.co.uk/artists-pages/in-the-name-of-progress/.

Fig. 4.4 Walking Middle Field on Lundgren's farm. *A Field of Wheat.* 24 June 2016. Photographer: Susan Haedicke

In the second event on the farm on 24 June 2016, participants revisited the field once wheat had grown waist-high to engage in embodied learning about the farm's biodiversity and the water pipes and drainage system under the soil. A discussion at the end of the day enabled the participants to ask questions and reflect on their experiences[41] (Fig. 4.4).

The final two-day event on the farm in September was around the harvest that had occurred a few days earlier.[42] Culhane and Levene framed the event as an invitation to the Collective to co-produce this finale of the project by sharing their skills and practices within the structure of the

[41] For photographs, see 'News and Events', http://fieldofwheat.co.uk/newsandevents/.

[42] For photographs, see 'News and Events', http://fieldofwheat.co.uk/newsandevents/. For a description, see 'Farmer's Almanac: It's in the barn!' http://fieldofwheat.co.uk/2016/08/20/the-wheat-is-in-the-barn/.

weekend of talks and celebrations of the harvest.[43] On Middle Field, for example, two members of the Collective, architects from Studio Polpo, created a special temporary straw bale structure for group discussions, meals and film projections. The core of the first day was a face-to-face version of a 'Critical Inquiry'—a participatory online dialogue that I will discuss later in the chapter. The second day of the Harvest Event featured workshops, discussions, games, performances and installations all around the farm and in the village hall where collective members had set up stalls of activities, such as painting with soil pigment, bagging the flour ground from the harvested wheat, collective baking with the flour, making art objects with the wheat, creating musical instruments from detritus from the farm and making book covers from fabric that had been buried in the field earlier in the year to dye it. A group of actors, the 'plough-players', burst into this fairground type of setting with their Plough Play, a twelve-minute Commedia dell'arte type skit, that had been performed in the village since 1908. Later, on the farm, a performative adaptation of an old harvest ritual took place.[44] Culhane explained that over the centuries, the potency of the last sheaf of wheat to be harvested, to 'come home', had played an important role in harvest rituals. This last sheaf was believed to harbour the spirit of the corn god, and many were frightened of its potency so they would throw a sickle at it to cut it. The sheaf would then be gathered and transformed into a corn dolly, a mask or some other art object. In the spring, it would be planted into the ground or burned. Culhane created a corresponding ritual[45] where she slowly entered the field wearing a large mask made of wheat[46] and walked to the final patch of wheat not yet harvested. She sank into the wheat and removed the mask. A volunteer took the mask and processed with it around the field. While this was taking place, another performer, wearing a headdress made from swan feathers that Culhane had found in the field over a year earlier, stalked around its edges. She represented the wild spirit of the field in the form of swan. The ritual ended with the burning of the mask. A feast of bread and cake collectively made from flour milled from the wheat in

[43] Information about the artists thinking and design behind the harvest event came from an email exchange with the artists, 27 May 2020.

[44] 'Artists' Pages: Middle Field Mask', http://fieldofwheat.co.uk/artists-pages/middle-field-mask/.

[45] See photographs, http://fieldofwheat.co.uk/newsandevents/.

[46] Culhane in this mask is the image on the book cover.

Middle Field closed the day. The conversations and activities around this performed ritual, including the Alternative Ploughing Match mentioned earlier, revealed how the dialogism of the project not only empowered the Collective to take responsibility for much of the harvest event, but encouraged them to participate in the agri-food system at least for the duration of the project.

While the face-to-face events promoted dialogue, knowledge exchange and comradery, the truly innovative, performative threads in *A Field of Wheat* occurred on the project's extensive interactive website with its six sections each offering different storylines for creative knowledge acquisition, engaged listening and dialogic interactions. The website was the Collective's primary encounter with the farmscape, especially for those who lived abroad and never visited the farm, and it was actually the *local place* where many of the project's activities occurred, so respectful conversation across difference was a key goal. Anne-Marie Culhane explains:

> In *A Field of Wheat* what we want to explore is twofold: designing the spaciousness of silent and attentive listening into the structures we were creating for online dialogue and holding a level platform for sharing different perspectives and viewpoints and levels of experience in ways that honour diversity and develop our collective intelligence.[47]

But these conversations are not easy in a project like this, Levene explains, 'where everything is so complex, so opinionated, so heated, you can lose touch by getting theoretical, argumentative, technical, romantic, emotive' (Personal Interview, 19 March 2019). So the artists chose to tie a thread of etiquette based on Quaker principles of respectful listening in Quaker worship into the project's knots.[48] This thread was essential in all the dialogic conversations, but particularly those online, and the artists worked with Quaker elders and Collective members to adapt the principles designed for face-to-face gatherings of worshippers to an online arts and agriculture conversation. 'The Exchange: Collective Inquiries

[47] 'Artists' Pages: Spaces for Listening', http://fieldofwheat.co.uk/artists-pages/spaces-for-listening/.

[48] See 'Artists' Pages: Spaces for Listening' for entries from both Culhane and Levene on the decisions around relying on Quaker principles and links to images of the meeting house that became part of the project, http://fieldofwheat.co.uk/artists-pages/spaces-for-listening/ and see also Joy, n.d., an article on Quaker practice.

and Decisions' was where much of the online dialogue and 'performativity' occurred. It created a unique farmscape beyond the farm that relied on two types of dialogue: communal reflection on philosophical ideas related to farming in the 'Collective Inquiries' and communal debate and decision-making about specific tasks in relation to the growing and harvesting of the wheat in the 'Decisions'. Both the 'Collective Inquiries' and the 'Decisions' were based on Quaker dialogue that invited attentive listening, a space for silent meditation and respect for different opinions.

The first two 'Collective Inquiries' (in November and June) each offered a quotation to which the Collective members could respond only once and one by one in an online post up to 400 words in their chosen twenty-minute time slot. Levene and Culhane cast themselves as witnesses of the 'Collective Inquiries'. During the almost twenty hours when members of the Collective posted their reflections one by one, Levene or Culhane was online waiting and reading the entries as they were submitted: a form of liveness, presence and listening, important in Quaker practice, but in a digital medium. For the first 'Collective Inquiry' in November 2015, just before the United Nations COP21 Climate Summit in Paris, Thomas Jefferson's quotation, 'While the farmer holds the title to the land, actually it belongs to all the people because civilization itself rests upon the soil'[49] was the prompt for the individual reflections. It was introduced with several facts about agriculture, notably that 40% of the earth's land surface is farmed and that 'climate change is both a product of and an influence on agricultural production'. The comments focused on meanings of and emotive reactions to the notions of ownership and stewardship of the land and the material agency of the complex ecosystem. While the responses varied, the farmscape that emerged in response to Jefferson's quotation, I felt, was closely aligned to Yi-Fu Tuan's notion of topophilia that foregrounded the farmer's strong bonds to the land he or she farms, a farmscape circumscribed by human/nonhuman activities, habits and partnerships that supported the need for stewardship of naturalcultural farmscapes. The second 'Collective Inquiry' in June pushed some of the ideas tentatively suggested in the first Collective Inquiry into focus. Inspired by Felix Guattari's words in *The Three Ecologies*: 'Now more than ever, nature cannot be separated from culture',[50] the

[49] See http://fieldofwheat.co.uk/collective-enquiries/ce-1/.

[50] See http://fieldofwheat.co.uk/collective-enquiries/collective-enquiry-no-2/.

members of the Collective grappled with the significance of understanding culture and nature as inseparable, particularly in agriculture. Lundgren offered a thought on the differences between agribusiness and agriculture: 'Agribusiness is producing food to the lowest unit cost so that others in the food chain can add value; agriculture is producing safe wholesome food whilst offering the producer a fair return on time and investment, enhancing the environment and supporting rural economy'.[51] He located his farming practices on the agricultural side of the continuum and no one challenged the inherent conflicts in using chemical pesticide and fertiliser and 'enhancing the environment', in respect, I think, to the pledge to follow Quaker principles. Inevitable tensions like this one often led to conversations outside the 'Collective Inquiry'.

In another line of conversation, Collective member, Carol Farrow, expanded the concept of culture:

> Culture doesn't just make me think about 'the Arts.' It makes me think about the natural yeast and enzymes in sourdough bread, Kombucha and ginger beer 'cultures'. Such foods feed the bacteria in our guts. Like the soil, they are life giving.[52]

John Baxter, the 'ponderer' appointed to 'distil' the insights in the 'Collective Inquiry', identified the farmscape emerging from the responses as one dependent on a form of listening and care that seemed to follow the line of thinking in Haraway's 'naturalcultural assemblages' that require a *response-ability* and a partnership 'to sow worlds with' (Haraway 2016: 125). Baxter related Carol Farrow's expansion of the idea of culture as 'a "collective stomach" digesting what the world offers and doing the work of renewal' to Lucy Neal's 'development of this image in her description of artists as earthworms "tilling the soil, preparing the ground to grow new perspectives, seeing and nurturing different ways of framing how we look at and take part in the world"'.[53]

'Seeds for the Future', the final 'Collective Inquiry', in September 2016 (mentioned earlier) was face-to-face in the Brant Broughton Quaker Meeting House that had inspired the use of Quaker tenets and values in this project. Here, the provocation was not a quotation, but rather a

[51] See http://fieldofwheat.co.uk/collective-enquiries/collective-enquiry-no-2/.

[52] See http://fieldofwheat.co.uk/collective-enquiries/collective-enquiry-no-2/.

[53] See http://fieldofwheat.co.uk/collective-enquiries/collective-enquiry-no-2/.

request to think about the future of farming and food through lenses of the project's storylines and the participant's own storyline about the field. The varied responses resonated with the insights into farmscapes in earlier 'Collective Inquiries' as the participants mused about food preparation and consumption, quality of food, narrowing distances between the general public and farming and the intricacies and journeys in the project and its participation in wheat-growing.[54] The 'Collective Inquiries', structured to imitate an actual Quaker meeting (Quaker services are called 'meetings'), all followed an etiquette where a comment made aloud was not a prompt to initiate debate, but more a thought on which others could reflect. The 'Collective Inquiries', for me, were about listening and contemplating rather than discussion.

For a dialogic space, the Collective relied on 'Decisions' that also took place online. In contrast to 'Collective Inquiries', 'Decisions' provided a space for debate and disagreements that could lead to new ways of thinking. Writing about radical democracy, Mouffe claims that conflicts must be aired since their erasure signifies an imposed agreement as in an authoritarian state. Tensions should not be silenced, but rather be allowed to contaminate ideas and individuals 'in the sense that…each of them changes the identity of the other' (Mouffe 2000: 10). Agonism, for Mouffe, is:

> A relation not between enemies but between 'adversaries', adversaries being defined in a paradoxical way as 'friendly enemies', that is persons who are friends because they share a common symbolic space but also enemies because they want to organize this common symbolic space in a different way. (Mouffe 2000: 13)

The 'Decisions' placed the members of the Collective in a similar agonistic space that sought to 'harness [conflict] in a productive way' (Mouffe 2000: 9). As the 'Decisions' placed the members of the Collective in the role of decision-makers about specific farming approaches to Middle Field,[55] the decision had to be debated, voted on and then enacted, even

[54] For the full audio recording of this Collective Inquiry, see http://fieldofwheat.co.uk/collective-enquiry-3/.

[55] All appear in full in 'The Exchange: Collective Inquiries and Decisions', http://fieldofwheat.co.uk/the-exchange/.

if not all the members were in agreement. Lundgren framed the decision process with his detailed knowledge of the field, but he agreed to follow the Collective's decision for action. The project had planned to allow the Collective to make three decisions, but the final one in July that asked about what to do with the straw had to be cancelled since the only viable option at that time for the field was to chop the straw and return it to the soil to improve its potassium deficiency. The other two decisions went ahead as planned. Each set of 'Decisions' had three stages carefully choreographed by the artists. The first stage was deliberately broad as it explained issues, gave information needed to take the decision and invited discussion. The second stage grew out of the dialogue in the first stage and focused on specific actions that needed to be done in this field of wheat. Its intended result was in a limited list of options. The third stage was the 'final voting' on which option to follow.[56] 'Decision One' in February asked the Collective to decide about the application of nitrogen fertilizer. Stage One of the decision process provided wide-ranging background information on nitrogen fertilizers and Lundgren's thoughts about its use on his farm. The subsequent dialogue among the Collective members was extensive and detailed, and it revealed opposing views, particularly around organic vs. non-organic farming methods and the role of the farm in the context of the agri-food system for both the farmer and the consumer. Not surprisingly, since this decision was about the application of a chemical fertilizer, the tensions in the collective decision-making process were clearest in the discussions around commercial vs. organic farming. Several members looked back and questioned the choice of wheat (a feed wheat, JB Diego, rather than a wheat for organic bread-making), but Lundgren reminded the Collective that the choice of wheat was linked to a key goal of the project: 'to help the collective explore the relationship between an individual field of wheat in Lincolnshire and the anonymous global food market'.[57] Discussion revolved around a few key issues: the impact of nitrogen fertiliser on the environment and its widespread use in conventional farming, the need to make a living from wheat production, and the cost and value of food to our lives. Given the wheat variety and the detailed information explaining

[56] Information about the three stages of the decision process came from an email exchange with the artists, 27 May 2020.

[57] See 'The Exchange: Decision No.1', http://fieldofwheat.co.uk/decisions/decisionno1/.

that it was designed for high yield through fertilizers, the Collective reluctantly began to understand why the option of no nitrogen fertilizer for the whole field was not really viable if Lundgren's economic sustainability was added to the decision process.[58] In the second stage of the 'Decision One', the discussion revolved around what to do about nitrogen fertiliser in this field that allowed little experimentation. The possible options discussed were replacing nitrogen fertiliser with an organic alternative like compost tea, reducing the amount of nitrogen fertiliser, or splitting the field so that part received nitrogen fertiliser and part used less nitrogen or only organic alternatives. In the summary of Stage Two, the artists wrote:

> In order to explore, understand and observe the complexities of modern wheat farming in this country and unravel its networks and impacts in the global context, we have had to start somewhere. Conventional farming is the reality of 'where we are now' in UK agriculture. Observing and being implicated in a growing cycle is a good starting point to see and feel how this system works'.[59]

After the period for comments (Stage One) and the extensive discussion of options (Stage Two) closed, the final decision, based on the number of votes for a particular path, was taken to leave a small part of the field (100 square meters) nitrogen-free in order to test other organic fertilising strategies and to see if the yield was different in that part of the field. (The Collective, of course, understood that this bit of the field would not be completely nitrogen-free.) The rest of the field would receive reduced levels of nitrogen. When the wheat was harvested and officially evaluated, the results showed that the wheat in both parts of the field had a high protein content so the reduced nitrogen levels did not have a detrimental effect. Those results, coming from the Collective's decision that Lundgren would not have made on his own, convinced him to continue experimenting with reduced nitrogen levels and organic fertilisers in subsequent years.

'Decision Two' (May to July 2016) focused on what to do with the wheat after harvesting: to sell the wheat on the futures market, to sell

[58] 'The Exchange: Decision No. 1', http://fieldofwheat.co.uk/decisions/decisionno1/.

[59] 'The Exchange. Decision No.1. Stage 2', http://fieldofwheat.co.uk/decisions/decision-no-1stagetwo/.

the wheat at the time of harvest to a farmers' co-operative or a global grain trader or to discover some other more creative option. The dialogue around 'Decision Two' encouraged engagement with the larger context of supply chains in wheat production. As part of the overall design of this decision, the discussion was directly informed by the Field Trip to London and the information shared on that day. 'Decision Two' was less emotive than the first one, although it did have an impact on profit levels, so the discussions were much less extensive. The decision taken was to sell most of the grain to the farmers' co-operative retaining about one hundred kilos for milling and experimental purposes by the Collective. As a participant in these decisions, I felt the dialogue did address opposing positions but within a controlled frame, and the various stages demanded careful listening as much as active debate. The 'Decisions', and to a lesser extent the 'Collective Inquiries', revealed how difficult dialogue is in practice. In the hopes a civil discussion, the artists chose to adapt Quaker tenets, and while that choice minimised conflict, it may have reduced new insights and discoveries. But I think the real goal of these exchanges was to begin training the members of the Collective to be meaningful participants in the agri-food system as food citizens. This sense of urgency to participate in the farm's production and distribution of the wheat was evident in the other sections of the website as well.

In 'Farmer's Almanac', Lundgren detailed the field's annual wheat cycle with vivid descriptions, thoughtful musings, photographs and information about the field of wheat and the farming activities to ensure a healthy crop in postings written by Lundgren. Collective members could ask Lundgren questions after each post and challenge his answers. The posts, questions and responses in this dialogic section of the website revealed a profound willingness to engage with knowledge exchange as Lundgren walked non-farmers through what it means to grow the wheat, including contentious issues around pesticides and fertilisers, and the Collective members contested long-held beliefs about farming practices. The 'Artists' Pages' shared information about the research that went into the project, for example, about the fens slodgers that foraged the fens before they were drained, the wheat futures market, and Quaker principles. Culhane and Levene also offered artistic responses to the developing project, such as following the wheat as it grew in drawings and photographs. And, the section of the website called 'Local/Global' placed the local Lincolnshire project in a global context. Each day, this page gave (and continues to give) the weather and current temperature on Middle

Field and the price of wheat on the global market for that day as well as the previous day, month, year, three years and five years. Another type of dialogue occurred in the one section of the website that was private to members of the Collective. Here members could offer contributions like bread recipes, links to radio programmes or Collective member's projects such as predicting yields or testing the soil. This section used a relational aesthetic that grounded the project in social interactions that relied on the members creating a space of knowledge exchange. The website created a vibrant virtual farmscape that balanced both local and global agricultural concerns and the needs of a working farm and a lived farmscape.

A Field of Wheat warranted a longer 'thick description' than some of the other projects not only because of its innovative performative practices, but also because all the dimensions of a food democracy, as identified by Hassanein (Hassanein 2008, discussed in Chapter 1) were woven into its structure. Conversations and knowledge acquisition/exchange were core practices of all aspects of *A Field of Wheat*, and the participatory nature of the face-to-face events and the online activities led to a strong sense of collaboration or networks among the members of the Collective, the artists, the farmer and the field of wheat. We were in the wheat growing and harvesting together. Hassanein also claims that in a food democracy, citizens place community good over their own self-interests (Hassanein 2008: 291). This care for others, whether the immediate community, a wider agricultural community or the field itself and, by extension, farmlands permeates all aspects of the project from its application of Quaker tenets to the discussions to dialogue around stewardship of the land. The other projects in this chapter shared those participatory practices with *A Field of Wheat*. But this project wove decision-making into its practices as well. Hassanein argues that a food democracy must provide individuals with efficacy or the ability to makes decisions that can produce alternative results (2008: 290). In 'The Exchange: Decisions' on the website, the Collective was given that efficacy and the knowledge to make informed decisions that could benefit the farm. And what we decided was what happened. Other areas of the project, like the sharing of skills during the harvest event, the sharing of reflections in the 'Critical Inquiries' and the sharing of recipes, projects and readings in the section of the website specifically for the members also empowered the Collective.

The complex dramaturgical structure of *A Field of Wheat* was diagrammed in a chart of the durational event's timeline that illustrated how *A Field of Wheat* performed an annual agricultural cycle of planting, growing and harvesting overlaid by a cycle of performative activities relating to the seasons with almost mathematical precision. It consisted of five concentric circles. The inner most circle stated the title and dates of the project; the second circle laid out the months with September marking the first month of the project. The third and fourth circles mapped the 'Farmer Activity' with blue dots indicating when certain agricultural tasks are done from the planting of the seeds at the end of September, followed by treatments of herbicides, pesticides and fertiliser from January through July to the harvest in late August or September as well as the crop growing cycle. The outer circle mapped the activities of the Collective, both the online Collective Inquiries and Decisions and the face-to-face events onto the activities happening in the field. I think that this diagram not only presents the varied activities in this project, but that it can also act as an iconic visual representation of the complexity of multiple storylines and tangled processes of a farmscape (Fig. 4.5).

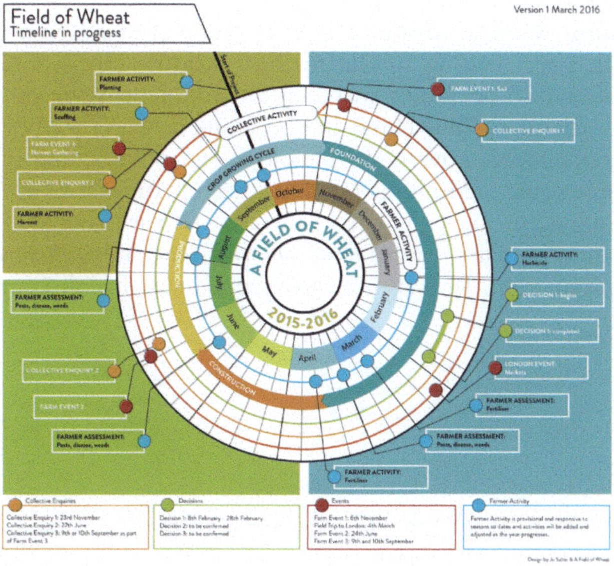

Fig. 4.5 'Field of Wheat Timeline' created by Ruth Levene and Anne-Marie Culhane with Jo Salter. *A Field of Wheat*. 2015–2016

The project and, of course, the field continue to have a life.[60] The harvested wheat was sold at a profit.[61] At the end of September 2016, Peter Lundgren planted a scroll that 'depicts the insects who live on Middle Field, painted in earth pigment made from soil samples taken from the field. It holds hand-written wishes for the future of the field and was co-created by collective members', as he wrote in his final post in the Farmer's Almanac.[62] This 'planting' echoed the replanting of the last sheaf of wheat in the old harvest festivals. This complex and durational performative project mapped a somatic experience that wove together walking, sensing, listening, speaking, sharing and empowering onto an actual and a virtual farmscape to enable non-farmers to understand industrial-scale wheat farming in the UK and farmers to hear from the public about their wheat first-hand.

Through the dialogic aesthetics of the farmscapes beyond the farm in this chapter, the artists not only raised audiences' awareness around pressing agricultural issues, but also created an actual and virtual dialogue to hear, understand, respect and respond to differences. In today's polarised world, such collaboration is urgently needed. In a piece in *The Guardian*, George Monbiot argued that Covid-19 compels us to recognise that we 'belong to the material world' (Monbiot 2020: 2), that care of and co-operation with our environment is the only way to assure our survival. '[T]his could be the moment when we begin to see ourselves, once more, governed by biology and physics, and dependent on a habitable planet' (Monbiot 2020: 2). Each performative event in this chapter, while occurring years before the coronavirus pandemic, revolved around farmer/nonfarmer/farmland *conversations* and *collaborations* to link food production, land use, environmental sustainability and citizen engagement.

[60] See http://fieldofwheat.co.uk/newsandevents/.

[61] See http://fieldofwheat.co.uk/2016/08/31/the-results-are-in/.

[62] See the text of the post and photographs, http://fieldofwheat.co.uk/2016/09/30/the-future-of-the-field/.

Works Cited

A Field of Wheat: Arts and Agricultural Project. n.d. http://fieldofwheat.co.uk/.

Beuys, Joseph. 2006. 'I Am Searching for Field Character, 1973'. In *Participation*. Ed. Claire Bishop. London: Whitechapel. 125–126.

Bishop, Claire. 2004. 'Antagonism and Relational Aesthetics'. *October* 110: 51–79.

Bishop, Claire. 2006. 'The Social Turn: Collaboration and Its Discontents'. *Artforum* February: 178–183. https://www.artforum.com/print/200602/the-social-turn-collaboration-and-its-discontents-10274.

Bishop, Claire. 2009. 'The Social Turn: Collaboration and Its Discontents'. In Rediscovering Aesthetics: Transdisciplinary Voices from Art History, Philosophy, and Art Practice (Cultural Memory in the Present). Eds. Francis Halsall, Julia Jansen, and Tony O'Connor. Stanford: Stanford University Press. 238–255.

Bishop, Claire. 2012. *Artificial Hells: Participatory Art and the Politics of Spectatorship*. London and New York: Verso.

Bourriaud, Nicolas. 2002. *Relational Aesthetics*. Trans. Simon Pleasance and Fronza Woods. Dijon-Quetigny: Les Presses de reel.

Braham, Emily. 2016. ''Milking Art to Improve Food Ethics'. *Positive News*. https://www.positive.news/environment/milking-art-to-improve-food-ethics/.

Cook, Ian. 2000. 'Social Sculpture and Connective Aesthetics: Shelley Sacks's *Exchange Values*'. *Ecumeme* 7.3: 337–343.

Culhane, Ann-Marie. Personal interview. 1 March 2019.

Denes, Agnes. n.d. *Wheatfield: a Confrontation*. http://www.agnesdenesstudio.com/works7.html.

Deutsche, Rosalyn. 1996. *Evictions: Art and Spatial Politics*. Cambridge, MA and London: The MIT Press.

Fischer-Lichte, Erika. 2008. *The Transformative Power of Performance: A New Aesthetics*. Trans. Saskya Iris Jain. London and New York: Routledge.

Freemantle, Chris. n.d. 'A Field of Wheat: Whose Art?' http://fieldofwheat.co.uk/wp-content/uploads/2016/09/A-Field-of-Wheatcfremantle.pdf.

Gonzalez, Robbie. 2013. 'Think you can walk in a straight line, blindfolded? Guess again.' *Gizmododo*. http://io9.com/think-you-can-walk-in-a-straight-line-blindfolded-gue-1387016277.

Haedicke, Susan. 2012. 'Opéra Pagaï's *Entreprise de Détournement*: Collage of Geographic, Imaginary and Discursive Spaces'. In *Theatre and the Politics of Space*. Eds. Erika Fischer-Lichte and Benjamin Wihstutz. New York: London and Routledge. 198–215.

Haedicke, Susan. 2016. 'Performing Farmscapes on Urban Streets'. *Popular Entertainment Studies* 7.1–2: 93–113.

Haedicke, Susan. 2018. 'The *Aroma-Home* Community Garden Project's Democratic Narratives: Embodied Memory-Stories of Planting and Cooking'. *Public Art Dialogue* 8.1: 114–130.

Haraway, Donna. 2016. *Staying with the Trouble: Making Kin in the Chthulucene*. Durham, NC and London: Duke University Press.

Harpin, Anna and Helen Nicholson, eds. 2017. *Performance and Participation: Practices, Audiences, Politics*. London: Palgrave.

Hassanein, Neva. 2008. 'Locating Food Democracy: Theoretical and Practical Ingredients'. *Journal of Hunger and Environmental Nutrition* 3. 2–3: 286–308.

Heim, Wallace. 2005. 'Navigating Voice'. In *Performing Nature: Explorations in Ecology and the Arts'*. Eds. Gabriella Giannachi and Nigel Stewart. Oxford and New York: Lang. 199–216.

Ingold, Tim. 2007. *Lines: A Brief History*. Routledge: London and New York.

Ingold, Tim. 2015. *The Life of Lines*. Routledge: London and New York.

Jackson, Shannon. 2011. *Social Works: Performing Art, Supporting Publics*. New York and London: Routledge.

Jordan, Cara. 2013. 'The Evolution of Social Sculpture in the United States: Joseph Beuys and the Work of Suzanne Lacy and Rick Lowe'. *Public Art Dialogue* 3.2. 144–167.

Joy, Leonard. 'Collective Intelligence and Quaker Practice'. The Co-Intelligence Institute. http://www.co-intelligence.org/P-QuakerCI.html and http://fieldofwheat.co.uk/artists-pages/spaces-for-listening/.

Kavage, Sarah. n.d.a. *Industrial Harvest*. http://www.industrialharvest.com/.

Kagage, Sarah. n.d.b. 'Place-Based Artwork: *Industrial Harvest*. http://www.kavage.com/projects/industrial-harvest/.

Kester, Grant H. 2004. *Conversation Pieces: Community + Communication in Modern Art*. Los Angeles: University of California Press.

Kester, Grant H. 2011. *The One and the Many: Contemporary Collaborative Art in a Global Context*. Durham and London: Duke University Press.

Krulwich, Robert. November 22, 2010. 'A Mystery: Why Can't We Walk Straight?' NPR: Morning Edition. http://www.npr.org/blogs/krulwich/2011/06/01/131050832/a-mystery-why-can-t-we-walk-straight.

Lacy, Suzanne, ed. 1995. *Mapping the Terrain: New Genre Public Art*. Seattle, WA: Bay Press.

Levene, Ruth. n.d. *A Field of Wheat with Anne-Marie Culhane*. http://ruthlevene.co.uk/works/a-field-of-wheat/.

Levene, Ruth. Personal interview. 16 March 2019.

Monbiot, George. 25 March 2020. 'Covid-19 is Nature's Wake-Up Call to Complacent Civilisation'. *The Guardian*. https://www.theguardian.com/commentisfree/2020/mar/25/covid-19-is-natures-wake-up-call-to-complacent-civilisation.

Mouffe, Chantal. 2000. *The Democratic Paradox*. London and New York: Verso.
Phillips, Patricia C. 1992. 'Temporality and Public Art'. In *Critical Issues in Public Art: Content, Context, and Controversy*. Eds. Harriet Senie and Sally Webster. Washington and London: Smithsonian Institution Press. 295–304.
Phillips, Patricia C. 2003. 'Public Art: A Renewable Resource'. In *Urban Futures: Critical Commentaries on Shaping the City*. Eds. Malcolm Miles and Tim Hall. London and New York: Routledge. 122–134.
Reid, Nessie. 2015. R&D: Research questions (Set #1). https://themilkingparlourblog.wordpress.com/2015/08/12/r-d-research-questions-1/.
Reid, Nessie. 2016a. *The Milking Parlour*. https://themilkingparlour.org/.
Reid, Nessie. 2016b. *The Milking Parlour Film*. https://capefarewell.com/latest/projects/the-milking-parlour.html and https://themilkingparlour.org/.
Reid, Nessie. 2016c. *The Milking Parlour (Introduction Film)*. https://capefarewell.com/latest/projects/the-milking-parlour.html and https://vimeo.com/151214210.
Rojas, Laurie. 2010. 'Beuys' Concept of Social Sculpture and Relational Art Practices Today'. *Chicago Art Magazine* (November 29). http://chicagoartmagazine.com/2010/11/beuys%e2%80%99-concept-of-social-sculpture-and-relational-art-practices-today/.
Rose, Mitch. 2006. 'Gathering "Dreams of Presence": A Project for the Cultural Landscape'. *Environment and Planning D: Society and Space* 24: 537–554.
Ruscoe, Sybil. 15 May, 2016. *On Your Farm: The Milking Parlour*. BBC Radio 4. https://www.bbc.co.uk/programmes/b07b9x6p.
Sacks, Shelley. 2016. *Exchange Values*. http://exchange-values.org/.

CHAPTER 5

Exposing Hidden Migrant Farmscapes

From Field to Fork: Setting the Context

The landscape of large-scale industrial farms features enormous monoculture fields stretching to the horizon and vast industrial-scale sheds for the concentrated animal feeding operations (CAFO): what John Bowe calls 'the post-pastoral fields of industrialised modern agriculture' (Bowe 2003: 34) (Fig. 5.1). Here, the intensive farming methods of single-crop farms

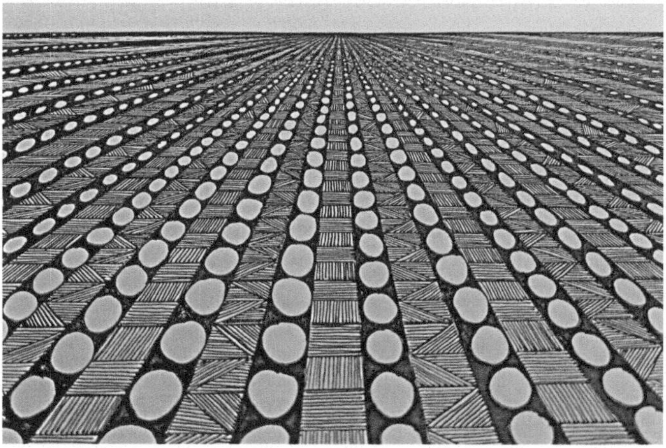

Fig. 5.1 *Impression Fallacies.* Pen and ink drawing by Kristen Oshyn

© The Author(s), under exclusive license to Springer Nature Switzerland AG 2021

S. C. Haedicke, *Performing Farmscapes*, Performing Landscapes,
https://doi.org/10.1007/978-3-030-82434-1_5

rely on extensive use of synthetic fertilisers and pesticides that are needed as the soil is depleted from growing the same crop in the same place year after year. Food production becomes an industry focused on producing as much food as possible, even if it goes to waste.[1] In *Dead Zone: Where the Wild Things Were*, Philip Lymbery introduced his indictment of industrial farms with:

> Planet-wise, the way we feed ourselves has become a dominant activity, affecting wildlife and the natural ecosystems that our existence depends on. Nearly half the world's usable land surface and most human water use is devoted to agriculture. Industrial agriculture—factory farming—is the most damaging. …Consciously or not, industrial farming has changed the way we think about food production. The system has switched from focusing on feeding people to producing more, regardless of whether it is consumed or not. More than half of the world's food now either rots, is dumped in landfills, or feeds imprisoned animals. (Lymbery 2018: xiv–xv)

The industrialised farmscape is one of power and profit on a large scale, and the experience of that landscape by the farm owners is often in the board room rather than on the land. There are exceptions, of course, as machines, such as huge combines, enable a single farmer to plant, cultivate and harvest thousands of acres of grain, and often these large-scale arable farms are family-owned and passed down from one generation to the next. As industry giants with economic and political clout, the owners of these huge farms have a powerful voice in global debates on food justice, labour relations and sustainability. Although these industrialised farmscapes perform very well in terms of increased yield, low-cost food for the consumer and low-cost grain for the animals in CAFOs, their performance in terms of the environment, biodiversity, animal welfare and human health is much less impressive. Since monocultures do not exist in the natural world, the human-engineered, ***un-natural*** monocultures have altered what these agrarian landscapes can support in terms of wildlife biodiversity and what they require to remain productive. As a result, the number of agricultural experts calling these farmscapes unsustainable

[1] See, for example, 'Advantages and Disadvantages of Intensive Farming' (2018), Balmford et al. (2018), Harum et al. (2013), 'Intensive Farming, n.d., Johnson (2017), Patel (2008), Shiva (2000), and Van de Zee (2017).

increases each year.[2] Their voices are not alone as activists and artists join the calls for concern and action with documentaries, political plays, eco-drama, community performances, agit-prop and performative protest art that accompanies direct action like strikes, marches, rallies, boycotts and public fasts. All this performance-based art operates at the intersection of politics and aesthetics to inform, engage and mobilise the public to be food citizens and to construct counter-hegemonic discourses. Performing these socio-economic and cultural industrialised farmscapes challenges the official agri-business narrative praising the benefits of intensive farming in terms of low cost and high yield with counter-narratives foregrounding the costs of that 'success' on the land, on democratic values of justice and dignity, and on the earth's ecology.

The United States leads the world in the number of industrial farms, but its model is rapidly being adopted around the globe (Harvey et al. 2017). The majority of American industrialised farms are either mega-farms (or 'factory farms') for livestock or mono-crop farms that produce enormous amounts of various grains, fruits or vegetables.[3] It is not surprising that this model developed in the American agricultural context since concepts and ideologies of space in the United States developed from the migration of Europeans to the New World who saw the continent as a vast landscape of opportunity, an 'empty' expanse of land full of natural resources 'waiting' for European expansion and 'improvement'. In the mid-nineteenth century, the phrase 'Manifest Destiny' that equated progress with taming the land was first used to describe this imperialist and expansionist push believed to be destined by God. American Studies scholar, Leo Marx, explained the American foundational myth about the land as being based on a utilitarian model: '[i]n this utilitarian interpretation of the myth, nature (… empty space, untapped resources, land uninhabited save for the savages) is a "howling wilderness." Its manifest destiny is to be discovered, subdued, and settled—made useful—by the arriving Europeans' (Marx 1991: 64). Since the land was there to be used for gain, it soon acquired 'the status of a commodity in a market or

[2] See, for example, Union of Concerned Scientists (2008).

[3] See McGreal (2019). Mary Swander's play, *Farmscape* (discussed in Chapter 2), addresses this issue that was a major factor in the Iowa farm crisis in the 1980s.

capitalist economy' (Marx 1991: 65). The longevity of the American utilitarian myth and the ways in which it plays out on many large-scale farms make the American context particularly helpful in understanding how these farmscapes perform in agriculture and how they are performed in the arts, but the issues they raise are certainly not limited to the American context.

As farms grew in size, the hired farmworkers became commodities as well, with the extreme form being the institutionalisation of slavery. In Rita Dove's play, *The Darker Face of the Earth* (1994), that adapts the Oedipus story to a pre-Civil War plantation in South Carolina, scenes alternate between the cotton fields and the 'big house' to expose the inequalities in a capitalist system that pits plantation owners against the commodified, enslaved workforce in this working farmscape and to dramatise resistance to that exploitative dehumanisation. Although Abraham Lincoln's 'Emancipation Proclamation' (1862) and the 13th Amendment to the Constitution (1865)[4] ended legalised slavery, the belief in land and labour as commodities was much harder to dispel, and it continues to dominate the ethic of vast farms today. While the huge green or gold fields of produce make stunning photographs, these images do not resemble the landscape experienced by those who work the land, particularly on large-scale farms producing fruits and vegetables for human consumption that, unlike cereals, require hand-picking to prevent bruising and so use large numbers of seasonal farmworkers at harvest. In the United States, these seasonal, often migrant, farmworkers are among the poorest, most exploited, and least protected workers in the country, with over 60% receiving an annual wage well below the poverty line.[5] The US Department of Labor's National Agricultural Workers Survey estimates that over 75% of those working on American farms are foreign-born, and up to 70% are undocumented ('Farmworker Justice' 2019).

[4] The Coalition of Immokalee Workers (discussed in Chapter 6) exposed several cases of modern-day slavery on farms in the southeast of the United States in the early twenty-first century.

[5] In her testimony before the United States Senate Committee on Health, Education, Labor and Pensions in 2008, Mary Bauer, Director of the Southern Poverty Law Center's Immigrant Justice Project, claims that 'Farm workers who migrate are poorer than settled seasonal labourers, with migrants earning $5000 per year'. With just a few exceptions (detailed in the work of the Coalition of Immokalee Workers in Chapter 6), little has changed in the years since 2008.

These startling statistics, however, do not support current claims that these illegal migrants are taking jobs from Americans. Many recent studies show that few Americans apply to do seasonal farm work. A compelling case study was done in North Carolina in 2011 when the North Carolina Growers Association listed 6,500 available jobs. Just 268 of the 489,000 registered unemployed North Carolinians applied. While 245 were hired, only 163 showed up on the first day. Only seven finished the season. Of the mostly Mexican workers who took the rest of the jobs, 90% worked to the end (Haspel 2017, n.p.). Without these often-invisible labourers, the US food system would not be able to function.[6] And yet, these crucial agricultural workers are routinely denied decent working conditions and basic human rights as the large numbers of undocumented workers not only fear deportation if they report abuse and forced labour, but also lack protection by US labour laws. For example, the Fair Labour Standards Act (FLSA), passed in 1938, banned child labour and guaranteed a minimum wage, but it did not include agricultural or domestic workers, jobs traditionally filled by African-American and immigrant populations (Canny 2005). Today, migrant farmworkers still struggle for basic human rights. The millions of migrant workers planting, cultivating and harvesting crops do not experience a bucolic landscape of beauty, plenty and prosperity; their experience is of a landscape of harmful pesticides, exploitative working conditions, hot sun, abuse and fear. While this chapter and the next focus on the plight of migrant farmworkers in the United States, exploitation of these essential workers is a global issue. It is this farmscape that activists and artists in this chapter and the next seek to expose and change.

Unlike frequent dramatisations of family farms, large-scale farmscapes only make a rare appearance on the theatrical stage. They are more likely to be found in prose literature[7] and on the street in a wide range of performative protest art. The farmer's sense of 'belonging' to and with the land, as evidenced in the case studies in the previous chapters, shifts to something more akin to a farm owner's belief in the equivalence of farmland and farmworker on the industrialised mono-crop fruit and

[6] See Wald (2011).

[7] Allison Carruth's *Global Appetites: American Power and the Literature of Food* (2013) looks at the link between both industrial agriculture (and its antagonists in the form of counter-cultural food movements) and US global power through a lens of twentieth-century American literature.

vegetable farms explored in this chapter and the next. Both land and labour are commodities to be used for corporate profit and personal gain. The story of the land is the story of the labourers as both are overworked and under-protected by intensive farming methods. The chemical fertilisers and pesticides and the industry model of increasing yield and profit over sustainable food production damage both land and labourer. These migrant farmscapes extend well beyond their role as spatial backdrop for traumatic stories of exploitation and abuse of farmworkers and rather become a lens through which to see the exploitation of labour and land required by today's global agri-food system and to understand the urgency for establishing food democracy. The performances explored here expose these hidden industrialised farmscapes, rife with abuse of both the farmlands and the farmworkers, and reveal those responsible for that negligent maltreatment. Performing these landscapes, whether in dramatic texts on the stage, documentary films on the screen or protest art on the streets,[8] constructs alternative narratives that strive to affect social and environmental change. These performances eschew ***about-ness*** in favour of ***event-ness***; they are less about ***showing*** or ***telling*** than about ***doing***. Here, artists and activists partner to theatricalise the migrant farmscapes and catapult them 'into the limelight' where they inevitably incriminate themselves. The artists and activists, and even the farmscapes themselves, become cultural commentators enabling the public to see what has been hidden. While this chapter and the next look at how artists present these American migrant farmscapes as experienced by the farmworkers and how they use art to create performative counternarratives to propel the public to demand justice, it is important to recognise that farmworker exploitation is a global issue and that protest art and performative activism is used worldwide to fight for their rights (Fig. 5.2).

[8] See sociologist Hank Johnston who claims that social movements appropriate the raw materials of the mainstream culture to create their cultures of resistance often articulated in the protest art that writes and disseminates the counternarratives. He categorises the 'raw materials' as ideations, artefacts and performance (2009: 7). See also Fahlenbrach and Scharloth (2016), Juris (2015), and Ratliff and Hall (2014).

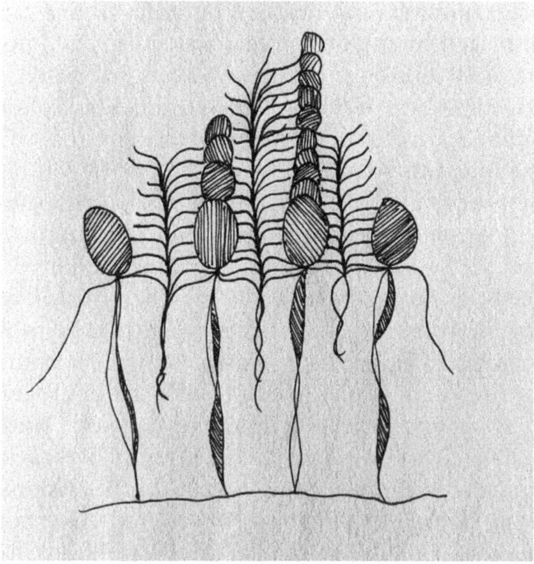

Fig. 5.2 *People lost for the fields*. Pen and ink drawing by Kristen Oshyn

The Legacy of John Steinbeck's *The Grapes of Wrath*

John Steinbeck's *The Grapes of Wrath*, published in 1939, inspired, directly or indirectly, the other case studies in this chapter. His epic novel drew on extensive ethnographic research on migrant farmworkers harvesting the crops in the fertile San Joaquin Valley in California that, in the mid-twentieth century, produced all the nation's raisins, 90% of the plums, two-thirds of the grapes and olives, and almost half of the peaches and nectarines—more agricultural goods than entire states except Iowa and Texas.[9] Steinbeck sought to raise public awareness of the living and working conditions of migrant farmworkers on these large-scale farms. His research on migrant labourers began in 1936 for a number of journalistic and creative projects, only one of which was published, a series of articles entitled 'The Harvest Gypsies' for the *San Francisco News*. Soon after 'The Harvest Gypsies', Steinbeck agreed to write a piece for *Life*

[9] See also, Mitchell (1996) that looks at migrant farm workers in California from 1913 to 1942.

magazine focusing on the California labour camps housing the migrants displaced by the Oklahoma dust storms and poverty. Crowded into substandard housing, the farmworkers harvested the crops they could not afford to buy. Working with documentary photographer Horace Bristol, Steinbeck travelled around these camps in 1937 and 1938. The misery of the migrants so moved Steinbeck that he felt he could raise public awareness more effectively in a novel, so he dropped the *Life* magazine project and wrote *The Grapes of Wrath* in five months. It was an instant success, and the protagonists, the Joad family, acquired a life beyond the pages of the book, even to the point that when Bristol, years later, finally published the photographs of the farmworkers that he took on the research trip with Steinbeck, he used the characters' names from the book rather than the farmworkers' actual names to title the photographs.

It is not surprising that Steppenwolf Theatre Company and Frank Galati chose to adapt *The Grapes of Wrath* in 1988 as the California migrant farmworkers were again in the national news. In 1982, Republican George Deukmejian was elected governor of California with more than $1 million in campaign contributions from growers, probably in response to the increasing success of recent United Farm Workers Union strikes. In 1984, Cesar Chavez, labour organiser, Latino civil rights activist and co-founder of the first farmworkers' union, called the third grape boycott and soon after initiated the 'Wrath of Grapes' campaign on the damage pesticides were wreaking on the land and the labourers.[10] And in 1988, Chavez began a thirty-six-day hunger strike to raise awareness of the hazards of pesticides. Frank Galati's adaptation of Steinbeck's *The Grapes of Wrath*, first performed at Chicago's Steppenwolf Theatre Company in 1988 and transferred to Broadway in 1990,[11] dramatised the story of the Joad family in Steinbeck's epic novel as a way to counter official narratives of abundance and fertility of the San Joaquin Valley with the poverty of the family.[12] It sought to reinsert the plight of the farmworkers not only into the agricultural landscape, but also into public discourse. Before examining how the case studies in this chapter responded, directly and indirectly, to Steinbeck's legacy, it is helpful to

[10] See United Farm Workers documentary, *Wrath of Grapes* (1986).

[11] See Frank Rich's review the New York production of *The Grapes of Wrath* that praises the production for not succumbing to 'sentimentality and cheap optimism' (1990: 3) and Jackson (2005).

[12] See Galati (2011) where he talks about his adaptation.

look at Jacques Rancière's linking of aesthetics and politics in his notion of collage that suggests a theoretical foundation to unpack the artists' dramaturgical strategies in the case studies in this chapter and the next.

Jacques Rancière and the Aesthetics-Politics of Collage

Rancière links aesthetics and politics through their 'ways of doing and making' (Rancière 2004: 23), through their participation in determining and often reconfiguring what is seen, heard and understood about the contemporary world. They are not linked through forms of representation, but instead both art and politics are activities that propose innovative perceptions and conceptions of current situations, challenge the status quo and seek ways to alter understanding so there is space to imagine new possibilities. Art and politics, for Rancière, are forms of dissensus that interrupt rules governing social experiences, present images and ideas that cut across normative beliefs, reconfigure the status quo by giving voice to those usually silenced and create spaces for new subjectivities. The efficacy of dissensus arises from this 'new landscape of the visible, the sayable and the do-able' (2010: 149) that imagines 'a new fabric of common experience, a new scenery of the visible, and a new dramaturgy of the intelligible' (Rancière 2010: 141).

Collage, for Rancière, acts like dissensus as it links aesthetics and politics through its reliance on a slippage between art and non-art that opens the door to thinking in contradictions: 'art is art to the extent that it is something else than art' (Rancière 2010: 118). The aesthetic-political logic of collage:

> mixes the strangeness of the aesthetic experience with the becoming-life of art and the becoming-art of ordinary life. Collage can be carried out as a pure encounter of heterogeneities, testifying wholesale to the incompatibility of two worlds. It's the surrealist encounter of the umbrella and the sewing machine, showing the absolute power of desire and dreams against the reality of the everyday world, but using its objects. Conversely, collage can be seen as evidence of the hidden link between two apparently opposed worlds. (Rancière 2006: 84)

Collage is necessarily political because its reassembling of ordinary objects and sites into new and often startling creations and spaces hovers at the spatio-temporal point of tension where a comfortable comprehension of a logical reality coexists with an unsettling disorientation caused by an alternative possibility as it calls unexamined norms into question. Similar to the Situationist strategy of *detournement* that changes or disrupts how one sees and understands familiar landscapes, collage reassembles, juxtaposes or superimposes incongruous images, ideas or logics to construct alternative worlds. Not surprisingly therefore, collage often 'takes the form of a shock, which reveals one world hidden beneath another' (Rancière 2006: 87).[13] Collage creates a disorientation where ordinary, recognisable objects or understandings are reassembled into new and startling creations. The familiar becomes unfamiliar as collage uses the quotidian reframed to seem strange, even menacing, but often liberating. The technique of collage can be discerned in the practices of the artists who expose hidden and silenced migrant farmscapes of industrial farms.[14]

The case studies in this chapter explore a range of political performance and protest art linked to social movements on controversial issues around large-scale agriculture. I argue that these works use techniques similar to Rancière's collage to cause aesthetic-political shifts that trouble the imaginaries of idyllic farmscapes and critique the vast monocultural crop fields in the American context. Rather than iconic agricultural landscapes of 'amber waves of grain' and 'fruited plain(s)' (immortalised in the lyrics of the 1895 patriotic hymn, 'America the Beautiful' and sung as part of a medley by Jennifer Lopez at President Joe Biden's inauguration in 2021), these case studies create a rupture not only by inserting the under-paid and under-protected seasonal labourers into a now-politicised farmscape, but also by equating the exploitation of the farmworkers with that of the land. Drawing on the concept of food democracy and its call for social

[13] See Claire Bishop (2004, 2009, 2012) where she explores shock and rupture as key aspects of the art experience.

[14] Rancière does add a cautionary note about the efficacy of art however: 'There is no straight path from the viewing of a spectacle to an understanding of the state of the world, and none from intellectual awareness to political action. Instead, this kind of shift implies a move from one given world to another in which capacities and incapacities, forms of tolerance and intolerance, are differently defined. What comes to pass is a process of dissociation, a rupture in the relationship between sense and sense, between what is seen and what is thought, and between what is thought and what is felt' (Rancière 2010: 143).

justice in the fields, each of the case studies enables the migrant farmscape to make visible its 'back-stage' reality to encourage the witnesses to imagine and demand alternative worlds of justice.

Harvest of Shame: 'This is an American Story—A 1960s *Grapes of Wrath*'

In the 1950s and 1960s, the working conditions in the fields of the large-scale monoculture fruit and vegetable farms had not improved from those described by Steinbeck in *Grapes of Wrath*. On Thanksgiving Day in 1960, *Harvest of Shame*[15] aired on American television.[16] This award-winning documentary, narrated by Edward R. Murrow, acted as a collage contrasting Thanksgiving as a traditional American holiday of feasting on the abundance of the harvest[17] with the horrific living and working conditions of white and African-American migrant farmworkers in the United States who picked fruits and vegetables up and down the East Coast, working in what Murrow called the 'sweatshops of the soil'. In his opening remarks, Murrow makes that juxtaposition clear. Standing in front of a field of neat crop rows that stretch into the distance and looking directly at his viewers, he claims that this story:

> has to do with the men, women and children who harvest the crops in this country of ours, the best fed nation on earth. These are the forgotten people, the under-protected, the under-educated, the under-clothed, the under-fed. We present this report on Thanksgiving because if it were not for the labor of the people you are going to meet, you might not starve, but your table would not be laden with all the luxuries we have come to regard as essentials. (*Harvest of Shame* 1960: n.p.)

Harvest of Shame vividly re-drew the idyllic image of the American agricultural landscape for millions of Americans who were exposed to images of the hot and dusty fields of produce sprayed with toxic chemicals,

[15] See *Harvest of Shame* (1960).

[16] Murrow says the documentary is presented on Thanksgiving Day (fourth Thursday in November), but some other sources place its airing on the day after Thanksgiving.

[17] The Thanksgiving holiday itself is a fraught collage of celebration of abundance and systemic oppression of Native Americans. The history of Thanksgiving and the debates surrounding it are outside the scope of this book.

crowded and dangerous buses used to transport farmworkers, and shacks or tents or simply blankets under the open skies to witness the lives of extreme poverty and hunger of these farmworkers. *Harvest of Shame* performed the industrialised farmscape through the stories of the migrant farmworkers who talked directly to the television viewer about living and working in that setting. The labourers were inserted into the farmscape certainly, but not just as statistics. Here, they had names, faces that showed their exhaustion and voices that expressed sorrow. The farmworkers were individualised and humanised as the viewers heard not only their hardships, but also their hopes for a better life. In Galati's adaptation of *The Grapes of Wrath*, Ma confides in her son, Tom, 'I like to think how nice it's gonna be, maybe, in California. Never cold. An' fruit ever' place, an' people just bein' in the nicest places, little white houses in among the orange trees. I wonder—that is, if we all get jobs an' all work—maybe we can get one of them little white houses' (Galati 1991: 22). Her words resonate with those of the migrant children in *Harvest of Shame* who told producer and interviewer David Lowe that they wanted to be teachers or doctors when they grew up. Later, when Lowe spoke to their teacher, she admitted that the children did not have much chance to fulfil their dreams since only one in five would finish grade school. Not one child of a migrant farmworker, Murrow told his television audience, had received a university degree. In fact, he continued, the US government spent $6.5 million to protect migratory birds, but refused to approve about half that amount to educate migratory children.

The documentary form enabled Murrow to present the disturbing information as a news story for which he was an eyewitness, and he used that sense of 'reality' to encourage his audience to engage in political action. But the documentary format in *Harvest of* Shame actually presents a more complicated and constructed version of the events. Film theorist Stella Bruzzi calls documentaries 'performative acts, inherently fluid and unstable' (Bruzzi 2006: 1) that represent a 'negotiation between filmmaker and reality' (Bruzzi 2006: 186), between 'the real event and its representation' (Bruzzi 2006: 13). Bruzzi argues that the goal of documentary is not authenticity, but rather accessibility: 'the raw material is incapable of drawing out or articulating the truths, motives, or underlying causes it both contains and implies, so it falls to the writer to extract this general framework' (Bruzzi 2006: 13). She claims that documentaries offer a representation of reality:

> a documentary will never be reality nor will it erase or invalidate that reality by being representational. Furthermore, the spectator is not in need of signposts and inverted commas to understand that a documentary is a negotiation between reality on the one hand and image, interpretation, and bias on the other. ...[A] documentary itself is the crucial point at which the factual event, the difficulties of representation and the act of watching a documentary are confronted. (Bruzzi 2006: 6–7)

Film theorist Vivian Sobchack argues that a 'documentary is less a *thing* than an *experience*' (Sobchack 1999: 241), and she shifts the focus from the authenticity of the documentary object to what she calls a '*documentary consciousness*' (Sobchack 1999: 241) that understands the experience of the documentary in terms of its effect on the viewer. The documentary itself becomes a performative site of debate, not in terms of its accuracy, but in terms of the consequences of the information and events it imparts. *Harvest of Shame*, through its use of documentary, creates a space for the industrialised farmscape to perform itself, to expose its underbelly of violence, poverty and injustice, but the documentary also acts as a lens through which to understand that it has multiple voices with shifting, often conflicting, points of view and tactics of collage that cause shock and encourage viewers to reflect and act on the material.

Harvest of Shame opens as the camera pans over the hundreds of men and women in an open parking lot bidding for a day's work harvesting in the fields. Murrow explains:

> This scene is not taking place in the Congo. It has nothing to do with Johannesburg or Cape Town. It is not Nyasaland or Nigeria. This is Florida. These are citizens of the United States, 1960. This is a shape-up for migrant workers. The hawkers are chanting the going piece rate at the various fields. This is the way the humans who harvest the food for the best-fed people in the world get hired. (*Harvest of Shame* 1960: n.p.)

Murrow then quotes a farmer who, on observing this daily ritual, interjects, 'We used to own our slaves; now we just rent them'. A bit later in the documentary, a chaplain, Minister Cassidy, who worked with the migrants, claims that the living and working conditions of the migrant workers 'are just as base as [that of nineteenth-century] slaves. Only in name, they are not a slave, but the way they are treated, they are worse than slaves, and somebody is making thousands of dollars out of his sweat. Is that a slave or not?' These images and voices offered a rarely-seen view of a landscape of industrial farmlands to the 1960s viewer. Gone are the iconic landscapes of a vast and plentiful agricultural paradise unsullied by dirty, sweaty labourers. Here is a landscape filled with people working hard under brutal conditions and cognizant that things are unlikely to change. Shifting the points of view between those hiring and those being hired, between a sympathetic eye and a callous one, complicates the reading of the morning event and engages the viewer as Murrow's authorial voice suggests that what they will see is 'real', but it will be unexpected and perhaps unbelievable. Murrow, in his role as narrator, uses his commentary to guide the audience's response through the maze of opposing perspectives, and he does not shy away from taking a strong position on the miserable conditions in which the farmworkers live and work. The juxtaposition of this unrecognisable farmscape of poverty, hunger, filth, abuse and exploitation with the feasts on the viewers' Thanksgiving tables created a collage similar to that described by Rancière, but one that juxtaposed a video image in the documentary with an image from the actual dinner table of the viewers, and it was meant to shock and disturb. 'This is an American story—it is a 1960s *Grapes of Wrath*', insists Murrow.

The stories told by the farmworkers to producer David Lowe expressed the poverty and hardship they faced in graphic detail as the camera showed the filth and lack of running water in the small rooms shared by whole families. Mrs Doby, a mother of nine, explained that everyone in the family works in the fields except for the baby and that the children must essentially take care of themselves. When Lowe asked how much milk they buy for nine children, she responded, 'Well, we don't — we don't have milk except maybe when we draw our pay check. We have milk about once a week'. Story after story of children fending for themselves, of families surviving on a pot of beans and sometimes sleeping in the open if the labour camp is too expensive, interrupt the excuses of the growers and politicians who insist that their hands are tied. Murrow's commentary suggests an aural form of collage as it braids the contradictions together, reveals the hidden story beneath the official one and makes sense of the images that so oppose the normative image of American farmlands in 1960. *Harvest of Shame* places the familiar landscape of abundance in juxtaposition to an unfamiliar one that exposes the farmworkers' plight. This Rancière-like collage highlights the absurdity of fields of plenty either empty of human activity or filled with happy labourers and causes a shift from perception of a geographic landscape viewed from a distance to a somatic experience of a political landscape provoking engagement. The collage forces an acceptance of a contradiction in order to understand a new reality. This strategy of juxtaposition of contrasts between mental images of agricultural plenty and actual photographs and film footage of agricultural poverty occurs throughout the documentary as the treatment of the workers is placed side-by-side with that of the livestock or produce. The workers crammed into dangerously crowded buses or standing in open trucks as they migrate from farm to farm contrast with the carefully packed fruits and vegetables shipped in refrigerated trucks or trains, and the journeys of migrants and cattle show how little care was taken for the farmworkers' well-being. The camera follows the migrants on the long journey from Maryland to Florida in buses that often drove a ten-hour stretch without a stop for food or facilities for up to four days and nights. The cattle on a similar journey were fed, watered and rested for five hours out of every twenty-eight hours. And the squalid conditions of the work-camps where the migrants lived for several weeks as they harvested a crop—families of six or more in one small room often sharing a bed, no hot water or electricity, few outhouses and a single outdoor faucet for the whole camp—sit in sharp contrast to the new expensive and spotless stables built for the horses.

Harvest of Shame is not subtle, but rather sought to disturb viewers with the brutality of the images, the indifference or worse of the employers, the reluctance of the politicians to pass legislation to relieve the poverty and the brutal hardships of the farmworkers. Murrow asks, 'Must the two to three million migrants who help feed fellow Americans work, travel and live under conditions that wrong the dignity of man?' The documentary sought to prick the conscience of a nation, and it offered a call to action by hoping to cause public outrage that could force legislation to redraw the industrial farm landscape not just by adding the workers to the landscapes to achieve some accuracy of the situation as it is, but rather to actually change the farmscape to one of safety and justice. Murrow closed with a call to arms (a call to become a food citizen):

> The migrants have no lobby. Only an enlightened, aroused and perhaps angered public opinion can do anything about the migrants. The people you have seen have the strength to harvest your fruit and vegetables. They

> do not have the strength to influence legislation. Maybe we do. Good night, and good luck. (*Harvest of Shame* 1960: n.p.)[18]

While small steps were taken after the documentary in terms of some funding for health services and education for migrant children, the poverty remained. Ten years later, in 1970, Chet Huntley of NBC revisited the migrant farmworkers and reported that little had changed: 'It's been ten years since Edward R. Murrow made *Harvest of Shame*. We hope no one will need to make a film about migrants ten years from now' (Blair 2014: 5). Yet additional documentaries were made in 1990, 1995 and 2010.[19] Clearly, public outrage at seeing the workers in industrial farm landscapes was not enough. Nevertheless, the transformation of anonymous farmworker 'commodities' into individual human beings with names and faces, goals and plans, created a strong condemnatory performance of these farmscapes. The documentary as 'experience' (Sobchack 1999: 241) became a site of resistance and a powerful tool for mobilisation in later decades.

Into the Fields: The United Farm Workers Union and El Teatro Campesino

A similar industrial farmscape of sub-poverty wages and horrific living and working conditions could be found in California in the 1960s as well—conditions no better than those described by Steinbeck.[20] In a recent interview, Delores Huerta, co-founder with Cesar Chavez of the United Farm Workers, remembers:

> [T]he conditions were terrible.... They didn't have toilets in the fields, they didn't have cold drinking water. They didn't have rest periods. People worked from sunup to sundown. It was really atrocious. And the families were so poor.... When I saw people in their homes—they had dirt floors. And the furniture was orange crates and carboard boxes. (Godoy 2017: 4)

[18] The documentary later inspired Stephen Colbert's comedic episode on farmworkers, also titled *Harvest of Shame*, in the Colbert Report in 2010. His subsequent testimony before Congress on the migrant farmworkers' plight complicated the concept of collage and political art seeking to affect social change. See Baym (2010: 209–230).

[19] See also Bullard (2015).

[20] See also Grossman (2002).

The Chavez family became migrant farm workers in California after the banks took their Arizona home in 1938 when Cesar Chavez was only eleven, so he experienced the poverty Huerta and Steinbeck described first-hand. In 1962, Chavez and Huerta joined forces and founded the National Farm Workers Association (NFWA).[21] In September 1965, the primarily Hispanic NFWA joined forces with the mostly Filipino Agricultural Workers Organizing Committee (AWOC) to call a strike against the Delano grape growers, and in 1966, NFWA and AWOC merged to become the United Farm Workers (UFW). From the beginning, Chavez and Huerta recognised the potential of performative protest to tell a counternarrative to the official one of success in increasing yield of low-cost food and contributing to the wealth of the area and, in so doing, to change public perception of industrial farm landscapes. As a result, the UFW worked simultaneously on political and aesthetic levels not only by placing performative protest art, like flags, banners and placards, alongside their union-organising activism, but also by constructing their outward-facing direct action, particularly long-distance marches, boycotts and hunger strikes (all meaningful tools to cultivate food citizenship) as performative events. The UFW relied on a strategy akin to collage's linking of aesthetics and politics where art and culture became the means to develop new modes of political action and new forms of social understanding. As Valdez explained in 'Notes on Chicano Theatre', the 'Huelga march to Sacramento in 1966 was pure guerrilla theatre. The red and black thunderbird flags of the UFEOC (then NFWA) and the standard of the Virgen de Guadaloupe challenged the bleak sterility of Highway 99' (Valdez 1994: loc. 115–120). Yet, he and Chavez recognised that the performative must work alongside, but distinct from, direct action and door-to-door campaigning to achieve social change. Rancière seems to echo that concern and cautions against a simple equation between aesthetics and politics. If the boundaries between them become too porous, he warns, they are in danger of cancelling each other out. While aesthetics and politics share dynamic characteristics and practices of rupture, instability and transformation, they do not function in the same way (Rancière 2010: 134–151). The long marches ('peregrinations' or pilgrimages as they were called) were often repeated over the next fifty years. Photographs and video showing worker solidarity in a

[21] See Pawel (2014).

quest for justice in the fields appeared in print media and on television and gradually began to change perceptions of the large-scale industrialised farmscape. The grape boycott (1965–1970), the first use of such a strategy in a major labour dispute in American history, not only succeeded in getting the growers to sign a union contract, it also had a significant impact on politicising the public. Thousands of supporters alongside the farmworkers organised the boycotts (United Farm Workers 2017). These allies and those who gave up buying California grapes during that period signalled a major step towards food citizenship.

In addition to the marches, Chavez conducted three fasts between 1968 and 1988 that contributed to the performance of the counternarrative. He chose to fast in 1968 to promote non-violence, and his hunger strike cast farmworkers on the side of justice and nonviolence and agribusiness on the side of injustice and violence. Each of the fasts had staged interventions, for example, in 1968, Senator Robert Kennedy joined 8000 farmworkers at a Catholic mass where Chavez broke his first fast. In 1988 when Chavez ended his longest fast of thirty-six days, the fast was picked up and continued for several weeks by religious and political leaders, like the Reverend Jesse Jackson and Kerry Kennedy (Robert Kennedy's daughter), and by celebrities, including Martin Sheen and Whoopi Goldberg. These performative public events, focusing on socio-political issues like the right to organise, fair wages, safe working conditions and the impact of toxic pesticides on farmworkers' health, created compelling counternarratives to the official narrative of the successes of industrialised farmscapes, and they influenced the dramaturgical strategies of later farmworkers' rights campaigns, like those led by the Coalition of Immokalee Workers (discussed in the next chapter).

An important addition to this art-activism paradigm was Luis Valdez's founding of El Teatro Campesino in 1965 on the Delano grape-workers' strike picket lines.[22] These early years of the partnership between El Teatro Campesino and what became the UFW, only lasting from 1965 to 1967, represent a unique period in both the union's and the theatre's history when El Teatro Campesino acted as the art-based outreach arm of the union. As Luis Valdez said in an early interview in 1967, 'We are dedicated to a very specific goal—the organization of farm workers.... The idea that really excites me ... is a theatre of political change.... I'm

[22] For addition information, see Broyes-Gonzalez (1994, 2006), Harding and Rosenthal (2006), and Huerta (1982, 2000).

talking about really influencing people, and I sense a hunger in art for this' (Bagby 1967: 79). During this time, Valdez, alongside the farmworkers, devised satirical skits or *actos* to be used as a tool by the union for organising and politicising. The *actos* told the same story of poverty and exploitation as *Harvest of Shame*, but in a very different form as they used humour to present the farmworker not as victim, but often as clever trickster, and they confirmed the power of the union with its united front demanding better working and living conditions. They were first performed on flatbed trucks on the 1965 picket lines and during the marches primarily to encourage the farmworkers to strike against the growers, but El Teatro Campesino soon began performing the *actos* beyond the fields to inform the public of the plight of the farmworkers and encourage them to boycott grapes.

For Valdez, the *acto* epitomised the reality of the life of the farmworkers: 'the *acto* is the social vision, as opposed to the individual artist or playwright's vision. *Actos* are not written; they are created collectively, through improvisation by a group. The reality reflected in the *acto* is thus a social reality' (Valdez 1994: loc. 182). The acto's apparent simplicity clarifies the socio-political context, and its satire humorously illuminates the power dynamics at play between the bosses and the farmworkers. Humour has long been recognised as important to political action since its 'critical role thus lies in poking a hole through often-undiscussed but official versions of everyday reality, exposing their contradictions and the arbitrary bases of their social power' (Paolucci and Richardson 2006: 334). The satiric *actos* effectively suggest a way to rewrite the industrial farm landscape in part through the cleverness of the farmworker, the underdog character (or *pelada/o*) who outsmarts the boss and overturns the ending of the story. In one famous *acto* from 1965, *Las Dos Caras del Patroncito* (The Two Faces of the Boss, Valdez 1994), the lowly farmworker, actually a hated scab who is forced to work or be deported, manages to overthrow the powerful farm owner by convincing him to try the 'carefree' life of a farmworker. The Patroncito opines on the 'good life' of the farmworker who receives free housing, transportation and food and has 'no worries,' at first sarcastically: 'All my Mexicans love to ride in trucks! Just the sight of them barrelling down the freeway makes my heart feel good; hands on their sombreros; hair flying in the wind; bouncing along happy as babies' (Valdez 1994: loc. 240). The farmworker vigorously shakes his head 'no' listening to the grower's words. The Patroncito's words echo those of one of the farm owners recorded in *Harvest of Shame*. When asked by Lowe if the migrant workers were happy, Jones responds:

> Well, I guess they got a little gypsy in their blood. They just like it. Lot of them wouldn't do anything else. Lot of them don't know any different. That's all they want to do. They love it. They love to go from place to place. They don't have a worry in the world. They're happier than we are. Today they eat. Tomorrow they don't worry about. They're the happiest race on earth. (*Harvest of Shame* 1960: n.p.)

Like Jones, the Patroncito begins to convince himself that his statements are true. The farmworker sees his opportunity and offers to let his boss try this carefree life for a day. By exchanging clothes and donning the Patroncito pig mask, the worker becomes the boss, and the boss, as farmworker scab, is taken away by his armed guard. The *pelada/o* character—the farmworker—is both victim and hero simultaneously and that contradiction is a source of power, similar to the strategy of collage described by Rancière.

By 1967, Valdez withdrew El Teatro Campesino from partnership with the United Farm Workers in order to expand the theatre's performances beyond farmworker issues and the needs of the incipient labour organisation. 'El Teatro Campesino was born in the huelga, but the very huelga would have killed it … When it became clear to us that the UFWCO would succeed and continue to grow, we felt it was time for us to move and begin speaking about things beyond the huelga' (Valdez 1994: loc. 140–145). It is particularly the theatrical work in these first two years when El Teatro Campesino was closely affiliated with the UFW and devised the political form of the *acto* that moved the performance of the industrialised farmscape to a more sophisticated level. El Teatro Campesino experimented with ways to bring aesthetics into political action and while these *actos* were quite didactic, they could increase farmworkers' awareness of the social construction of their exploitation. Theatre scholar, Yolanda Broyes-Gonzalez, identifies Mexican oral culture as a key source of inspiration for the *actos*.[23] Orality, she argues, is not just spoken word rather than written text; oral culture is words inextricable from their local context, the body that speaks them, and the physical memories of the community passed down in traditions, stories, ritual

[23] See Broyes-Gonzalez (1994, 2006).

and art. This oral tradition in its popular (or *carpa*) performance form relies on slapstick, stock characters (particularly the clever *pelada/o*) and satire to create counter-hegemonic performances that can make audiences laugh as they listen to difficult truths about their own cowardice, passivity or exploitation and that recognition can mobilise them to action. The blatant un-reality of the *actos* and the cleverness of the *pelada/o* character reinforce the idea that constructed situations can be reconstructed as a new reality even though the enacted solutions are absurd as in *The Two Faces of the Boss*. The style and purpose of El Teatro Campesino's *actos* to use humour to initiate reflection on the industrial farmscape in which they work reappear in the later skits (*teatros*) created by the Coalition of Immokalee Workers in their popular education sessions that inform farmworkers of their rights and challenge the way they understand their situation (Chapter 6).

Onto the Theatrical Stage: Octavio Solis' *Alicia's Miracle* and Cherríe Moraga's *Heroes and Saints*

Placing the industrial farm landscape on a theatrical stage, the next two case studies, Octavio Solis' *Alicia's Miracle* and Cherríe Moraga's *Heroes and Saints*, dramaturgically intensify the ways in which these farmscapes are performed through storytelling, imagery and sound inspired by actual events. Octavio Solis based his unpublished play, *Alicia's Miracle*,[24] on a devastating report from the Center for Investigative Reporting in 2014 entitled 'California's Strawberry Industry is Hooked on Dangerous Pesticides: The Dark Side of the Strawberry' that found that California's \$2.6 billion strawberry industry used pesticides called fumigants, 'linked to cancer, developmental problems and ozone depletion' (Yeung, et al. 2014: 1). Some of the chemicals used had been repurposed from World War I chemical weapons and so farmworkers, who were often working in or very close to the fields as they were sprayed, as well as local inhabitants, were at extreme risk. The investigators, Andrew Donahue and Bernice Yeung, found that Paul Helliker, a consultant for Dow AgroSciences and director of the California Department of Pesticide Regulation, created a loophole allowing Dow to develop a plan based on a 'cost–benefit calculus' that allowed growers to use pesticides above the acceptable

[24] Solis shared *Alicia's Miracle* with the author.

limit if the economic cost to the grower if the pesticide were banned was greater than the risk to farmworkers (Weisskopf 1988: n.p.). While this loophole was supposed to be temporary, in reality, Dow's 'new' plan allowed the loophole to remain open, and with Helliker's signature, strict oversight of pesticide use was dismantled and more than one hundred California communities were put at risk. The original 'risks and benefits' ratio was 'adjusted' towards the benefits side resulting in the cap for allowable toxic pesticides to be raised about sixty times over the original regulated cap that resulted in enormous profits for chemical companies and growers. Risks to farmworkers and residents were essentially brushed aside with demands for incontrovertible 'proof' that pesticides are the cause of illness. As Richard says to Alicia in *Alicia's Miracle*: 'Correlation does not equal causality' (Solis 2015: 18). At the start of the debate in 2001 that led to the report from the Center for Investigative Reporting, a toxicologist put her concerns about the increased cancer risk for these communities in writing. Nevertheless, corporate profit won out over environmental degradation and human health. Worst affected areas were in the profitable strawberry-growing counties of Ventura and Monterey. What was particularly insidious about this fumigant was that it did not leave residue on the fruit so the consumer was not affected, but its gases easily escaped into the air when it was pumped into the soil weeks before planting the fruit and so adversely affected workers and residents. One of the most problematic locations was around Oxnard's Rio Mesa High School, surrounded on all four sides by strawberry fields, and the neighbouring residential communities.

The Center for Investigative Reporting approached Tides Theatre in San Francisco asking them to dramatise the findings of this report, as well as others the Centre had conducted. Artistic Director Jennifer Welch took up the challenge and set up a programme called 'StoryWorks: The Tides' to foreground politically activist plays. The fourth play in the series was Donahue and Yeung's report on the strawberry industry. Octavio Solis was commissioned to write the play, and *Alicia's Miracle* was produced at Tides Theatre in 2015. It later toured in Spanish and English to the strawberry farming communities of Salinas and Oxnard (where the fictional Alicia lives and works) to inform them of the risks and encourage them to join with the Center for Investigative Reporting in the fight against the use of these toxic pesticides.

In *Alicia's Miracle*, the three positions in the debate on intensive use of toxic pesticides are embodied in three main characters: the toxicologist Hannah represents the voice of science trying to understand the toxicity of the pesticides to the people and the land and discern the 'measurable harm'; the Dow AgroSciences consultant Richard represents the powerful voice of the growers and chemical companies trying to preserve the profits gained from these intensive farming practices in spite of the probable health risks; and Mexican migrant Alicia Sandoval embodies the people living around the strawberry fields and affected by the chemicals. The character Alicia, pregnant with her first child, works in the cafeteria of Rio Mesa High School. While Alicia is a fictional character, Rio Mesa, as already mentioned, is the name of the actual high school in Oxnard, one of the most heavily sprayed agricultural areas in California, so the risks and benefits of the pesticides debate shifts from the theoretical to the personal as Solis blurs the boundaries between fact and fiction, between art and non-art, as in Rancière's collage. That dramaturgical strategy enhanced the play as an educative tool, especially in the local communities in which it was performed. The play opens just after a meeting on 'averaging' or permitting allowable amounts of pesticides on strawberry fields above the recommended level has concluded. Richard and Hannah disagree about the decisions made at the meeting. Richard says that the pesticides can be released for use on the fields since 'there's more than enough wiggle room to move ahead' and 'We've got to hustle or growers won't make their yield' (Solis 2015: 3). Hannah insists that the group decided to wait for more data on the carcinogenic impact. But before they leave, Alicia bursts into the room to confront them. She has just learned that her unborn child is deformed and dying. When asked why, the doctor explained, 'it's where you live, Mrs. Sandoval. It's where you work' (Solis 2015: 8): her child poisoned in utero by the pesticides she breathes every day. Armed with as much knowledge as she could find about the various pesticides online and her big knife, she wants to punish the individuals responsible for her baby's inevitable death. Alicia's intense physical pain caused by this poisoned pregnancy increases during the confrontation, and she begins to miscarry, but this loss imparts a clairvoyance into how the corporate greed and indifference to risks to those near the sprayed fields have entered her body: 'We don't matter! Not me, my baby, nobody! None of us matter!' (Solis 2015: 35). She realises that the fumigant Telone is the real father of her dying child, not her beloved husband: 'Can't you see? He's not your baby. He belongs to Telone. He's the baby of Telone. … [T]he chemical seed of Dow finds me, enters me, makes me its bride, plants its death inside me' (Solis 2015: 39). The chemicals have transformed her baby into 'the sick miracle of my life, the miracle inside me coming out now' (Solis 2015: 39). But this miracle child, presaged by her husband's dream that their 'baby would be special, that the boy would bring a message in the very act of being born, that he was a miracle' (2015: 24), will never be what they originally thought and hoped for. He is not a living child, but rather a dawning recognition that social change does not happen though a personal act of revenge, but through self-awareness and community resistance to systemic exploitation. Her baby's 'message in the very act of being born' is the urgent need to expose and vanquish the system that devalues human life in favour of profit and to organise active citizenship against it. It is an age-old story of the son overthrowing the father, only here the father is the pesticide, Telone.

Solis' play can be interpreted through a lens of performative application of the key tenets of food democracy. In its dramatisation of a detailed and damning investigative report, the play informed its audiences of the facts about the pesticides' impact on human health. And its merging of actual, recognisable locations like the Rio Mesa High School, familiar to the local inhabitants, and the fictional characters that stand in for the farmworkers, the scientists and the chemical corporations both personified the scientific debate and inserted it into a specific local context. But the play also acted as a rallying call to communities affected by the pesticides and to those opposed to corporate greed at the expense of social justice to join forces (to create coalitions in the terminology of food democracy) and fight the government policies that fail to reign in the chemical companies like Dow AgroSciences. Like *Harvest of Shame*, the performance-based work of El Teatro Campesino and *Alicia's Miracle*, the next case study, Moraga's *Heroes and Saints* seeks to inform and engage its audiences in political activism to create alternative farmscapes.

Cherríe Moraga's *Heroes and Saints*, produced by Brava! For Women in the Arts and first performed in 1992 at El Teatro Misión of San Francisco, adds an aesthetic complexity to the issues of risks of pesticides on the environment and people living and working there. Creating a play close to what Carol Martin calls 'theatre of the real' with its practices that 'recycle reality' (Martin 2013: 7), *Heroes and Saints* evocatively presents the assumption of the equivalence of the land and the labourers and the power of community direct action to change the farmscapes through both its reworkings of actual events and its adaptations of Chicano myths and symbols.[25] The land and the farmworkers are poisoned and dying, yet, in spite of the abuse, they find the strength to fight back and to begin a process of renewal. Moraga wrote *Heroes and Saints* in response to three events in the late 1980s. In 1985, a childhood cancer cluster was identified in McFarland, California, a town that called itself the 'Heartbeat of Agriculture'. Located in the San Joaquin farm belt, McFarland had reported sixteen cases of cancer in local children—eight times the expected number in a town of that size—and an unusual number of infant deaths, still births and low birth weights in the early 1980s. The 280-mile long valley used more chemical fertilisers and pesticides than any other farm region of equivalent size in the United States. As Michael Weisskopf explained to the general public in his 1988 article, 'Pesticides and Death amid Plenty': 'Pesticides are made to kill, and they leave a deadly trail. They hang in the foggy air of the valley, burrow into its loose, sandy soil, and contaminate underground drinking-water supplies' (Weisskopf 1988: n.p.). The valley was so saturated with cancer-causing chemicals that it was called a 'grand experiment' in the impact of these toxins on human health. The situation was allowed to continue even after studies showed that the pesticides caused birth defects and cancer, especially in children, because federal law protected the farmers by a 'cost–benefit calculus' (discussed in relation to Solis' *Alicia's Miracle*) that permitted pesticide use if risk to the farmworkers and the environment was lower than economic cost to the grower if pesticides were banned. The California politicians, elected in part through generous campaign contributions from the large growers and the agrichemical companies, listened to the powerful lobby in defence of pesticides and accused opponents of having little substantial evidence to link pesticides to cancers and of exploiting the tragedy of the cancer cluster to support their organisations (including the UFW and environmental groups). By the late 1980s, experts confirmed McFarland's cancer cluster but did not establish a direct link to pesticides. McFarland became the fictional town of McLaughlin in Moraga's play. The second event that inspired *Heroes and Saints* was the UFW's 1988 documentary, *The Wrath of Grapes* (United Farm Workers 1986) that blamed childhood cancers and birth defects on intensive pesticide use. One scene introduced Felipe Franco, born without arms and legs. His mother had worked in

[25] See Garza (2004), Greenberg (2009), Jacobs (2012), Thomas (2016) and Yarbro-Bejarano (2001).

the vineyards during her pregnancy. Felipe inspired the 'bodiless' character of Cerezita who appears as a head on a box, 'a rolling, table-like platform… her "raite"' (Moraga 1994: 90) that she operates with her chin. Cerezita is the product of toxins she ingested in the womb when her mother worked in the fields, but she is also metaphorically the human equivalent of the poisoned land—both the fields surrounding McLaughlin and the town itself built on a toxic waste dump.[26] Her body (or her lack of body) becomes the site of protest that enables the whole town to revolt and 'cleanse' the fields through fire. And the final historical event, the severe beating of Delores Huerta by the San Francisco police at a peaceful political protest in September 1988 became the dramatised beating of the character Amparo, a leader in the protests against pesticide use in McLaughlin (Moraga 1994: 89). In the play, the police attack on Amparo mobilises Cerezita to become the voice of protest and the symbol of justice. Like Solis, Moraga adapts the form of collage as she merges fact and fiction (art and non-art) to confirm that stories like the one in *Heroes and Saints* are also happening in actual local communities.

The play's startling wordless opening scene takes the audience into the vineyards that surround the town to witness the dramatic protest of the local children against the growers who spray the toxins poisoning the land and their town.

> *At rise in the distance, a group of children wearing calavera masks enters the grape vineyard. They carry a small, child-size cross which they erect quickly and exit, leaving its stark silhouetted image against the dawn's light. The barely distinguishable figure of a small child hangs from it. The child's hair and thin clothing flap in the wind. Moments pass. The wind subsides. The sound of squeaking wheels and a low, mechanical hum interrupt the silence.* CEREZITA *enters in shadow. She is transfixed by the image of the crucifixion. The sun suddenly explodes out of the horizon, bathing both the child and* CEREZITA. CEREZITA *is awesome and striking in the light. The crucified child glows, Christlike. The sound of a low-flying helicopter invades the silence. Its shadow passes over the field. Black out.* (Moraga 1994: 92)

Here, in this short theatrical scene, Moraga introduces the conflicting sides of the narrative, its key players and the provocative setting for the performance of the industrial farmscape and its opposing displays of

[26] Scholars, like Alison Kafer (2013) and Petra Kuppers (2007), encourage a rethinking of the linking of the disabled body to the natural world.

fertility and bounty, on the one hand, and death, on the other. The children wear calavera masks, the traditional skull masks central to El Día de los Muertos, a celebration that honours the deceased and acts as a reminder that the distinctions between life and death are not fixed as the dead are still an essential part of the communal life of the living.[27] The 1960s in the United States saw a revitalisation in Chicano communities of El Día de los Muertos traditions for political reasons. Drawing on strategies of collage, they became a way to honour the sacrifice of people lost through violence and injustice and to oppose capitalism. As artist and scholar Amalia Mesa-Bains claims the Mexican 'traditions around death and dying… have always been tied to resistance and struggle' (Hooks and Mesa-Bains 2018: 122). In the opening scene, the children plant the cross with the body of the latest child cancer casualty into the ground. By referencing El Día de los Muertos in this way, Moraga uses this powerful religious symbol of resurrection as a sacred link between the land and the migrant body, a link that suggests hope and salvation, and it simultaneously confirms that what the children are doing is an act of resistance. That theatricalised crucifixion, based on El Día de los Muertos rituals, not only establishes the responsibility for the deaths on the growers supported by US government agricultural policies, but also highlights the inherent contradiction in the industrial farm narrative glorifying intensive farming methods that rely on toxins to produce edible crops. In the published version of the play, Moraga poetically establishes the link between the religious reference to resurrection and the degradation of the land:

> *The hundreds of miles of soil that surround the lives of the Valley dwellers should not be confused with land. What was once land has become dirt, overworked dirt, overirrigated dirt, injected with deadly doses of chemicals and violated by every manner of ground- and back-breaking machinery. The people that worked the dirt do not call what was once the land their enemy. They remember what the land used to be and await its second coming.* (Moraga 1994: 91)

The presence of the wind so evocatively described in the staging instructions for the first wordless scene of the play (quoted in full earlier) adds

[27] For more information on El Día de los Muertos, see Brandes (1998), Hooks and Mesa-Bains (2018), and Marchi (2009).

to the desolation and despair of the event we just witnessed and encourages the audience to listen. The focus on sound is repeated as Cerezita is heard before she is seen and her slow arrival seems to draw the audience into the field with her. The carefully narrated staging of the opening scene locates the spectator as witness alongside Cerezita. As Diana Taylor explains, witnessing is not just about seeing; it is about being a participant in the event: a 'witness is part of the conflict and has a responsibility in reporting and remembering of events' (Taylor 1997: 25). Amparo, Cerzita's neighbour and one of the town's activists, explains the importance of witnessing to Ana Perez, the on-site newscaster who asks about the crucifixions, 'Why would someone be so cruel, to hang a child up like that? To steal him from his death bed?' Amparo responds, 'They always dead first. If you put the children in the ground, the world forgets about them. Who's gointu see them, buried in the dirt' (Moraga 1994: 94). The witness must enable the dead to become the voices of protest. The non-verbal opening scene ends with the intrusion of the growers, the faceless representative of corporate farming that seeks to keep the landscape of farmworker and land abuse hidden. The sound of a surveillance helicopter interjects an opposing landscape narrative to the one established up to this point as it ominously warns 'no witnesses'. Throughout the play, the growers have a menacing presence as the sound of low-flying crop-dusters and helicopters interrupts many of the scenes. The growers' awareness of the damage the pesticides are wreaking on land and labour becomes clear when they start spraying at night as Cerezita's brother explains: 'Nobody sees them that way. Nobody that matters anyway' (Moraga 1994: 122).

While the opening scene suggests Cerezita's role as witness to the injustice in McLaughlin, the rest of the play chronicles her transformation into 'hero' and 'saint' able to empower her town to oppose the power of the farm owners and US government. Her damaged body, poisoned by the chemical pesticides, parallels the poisoned land depleted by the intensive farming methods. Her body's and the land's 'lack' become powerful sites of protest as they shift the industrial farm narrative of farmworker's body and farmlands as useful commodities to a transgressive counter-narrative of rebellion that places care of the land and the workers above profit. Cerezita sees her lack as a formidable symbol for social change: 'If people could see me "ama", things would change,' she tells her mother (Moraga 1994: 113). So too with the farmscape—if people could *see* the damage, things would change is her transgressive message. She says to her mother, 'You tie my tongue, "ama". How can I heal without my tongue?' (Moraga 1994: 147) and when she is allowed to speak, she mobilises her community.[28] Understanding her symbolic power, Cerezita enlists the town's children to transform her into a vision of La Virgen de Guadaloupe, symbolic healer of the suffering of the Mexican people. In this form, Cerezita's words, physical presence and actions can renovate the image of her damaged body into a strong community one of resistance. Her empowered role as 'hero and saint', as leader of her community, defies her body's lack and instead becomes the source of a new beginning as Cerezita makes another pilgrimage to the field to plant the crucifix with another dead child in the poisoned ground in the final scene. The helicopter returns and interrupts the ritual by shooting Cerezita. Her death energises the town to revolt, and as they burn and so symbolically cleanse the fields, the play ends with 'the crackling of fire as a sharp red–orange glow spreads over the vineyard' (Moraga 1994: 149). In Moraga's *Watsonville: Some Place Not Here*, the sequel to *Heroes and Saints*, Juan, McLaughlin's priest and Cerezita's ally, describes what the growers did to Cerezita's body after shooting her in the final scene of the previous play. In an attempt to silence her even in death, they impaled her severed head on a thick grapevine post: 'They had forced the post through her mouth' (Moraga 2002: 69). But Cerezita could not be silenced as the display of her death revealed injustice and, Juan claims, she became the site of resistance and renewal: the new crucifixion. For him, this moment changed his understanding of the political situation and grower/farmworker power dynamics: 'And I understood— … How profoundly those men, with all their land and all their power, hated us. And I know that they would do anything, anything not to know their hate was fear' (Moraga 2002: 69). This idea is echoed in the actions of the growers that the Coalition of Immokalee Workers oppose (Chapter 6). Cerezita 'died for change' (Moraga 2002: 67), Juan insists, and Moraga, like Solis, suggests the agribusiness landscape can change through social action. Both of these plays dramatise local, working and naturalcultural farmscapes to propel audiences into action. And Solis' and Moraga's braiding of fact and fiction in presenting these multifaceted farmscapes heightens their potential for efficacy.

The case studies in this chapter have focused on how artists use and adapt strategies of collage, particularly juxtaposition of art and non-art

[28] This narrative trajectory follows the trope of the disabled body representing 'lack', but overcoming adversity through perseverance. See Kafer (2013: 141).

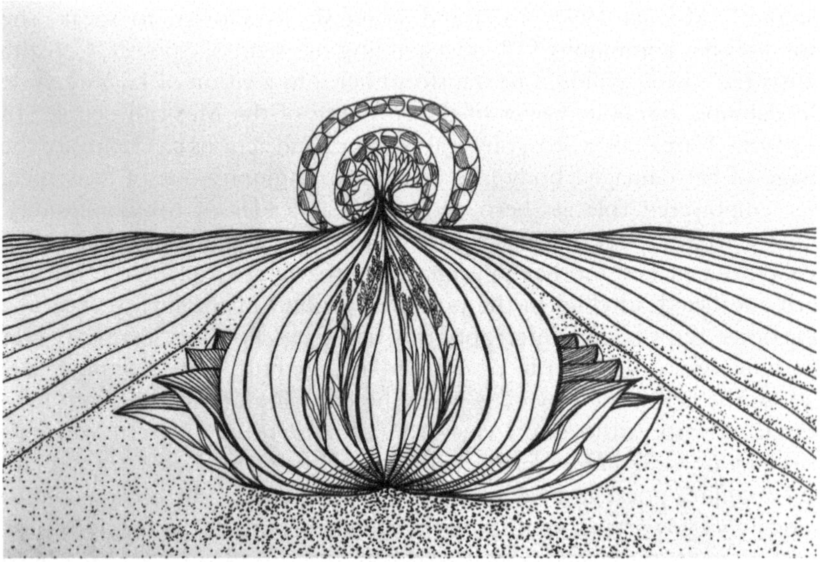

Fig. 5.3 *Support*. Pen and ink drawing by Kristen Oshyn

and blurring of aesthetics and politics, to both inform the public about the realities of large industrial farms and the plights of the farmworkers and inspire them to imagine alternative realities (Fig. 5.3). As in collage, the artists rely on creating shock as audiences see unexpected scenes of injustice firmly planted in actual local places. As discussed earlier in the chapter, Rancière claims that this shock causes a disruption in 'the relationship between the visible, the sayable, and the thinkable', a rupture that overturns expectations and 'resists signification' (Rancière 2004: 63). The experiential shock of this disorienting contradiction elicits an initial somatic response that then stimulates a desire to understand critically what had seemed natural moments before and is now unfamiliar. Grant Kester explained that 'we meet the epistemological challenge posed by aesthetic shock not by abandoning ourselves to the pleasures of ontic dislocation but by renewing, and expanding, our efforts to grasp the complexity of our surrounding world' (2004: 84). Shock is the way to encourage citizen activism in these case studies. The next chapter offers an extended case study on the efficacy of a partnership of direct action and protest art to transform the actual agri-business landscape as it looks at the human rights activism of The Coalition of Immokalee Workers that takes a step beyond shock and disturbing information as the farmworkers and their allies imagine and construct alternative migrant farmscapes based on justice and dignity and so affect social change by growing the number of food citizens.

Works Cited

'Advantages and Disadvantages of Intensive Farming'. 2018. *Farm Management*. http://www.farmmanagement.pro/advantages-and-disadvantages-of-intensive-farming/.

Bagby, Beth, and Luis Valdez. 1967. 'El Teatro Campesino Interviews with Luis Valdez'. *The Tulane Drama Review* 11.4: 70–80.

Balmford, Andrew, Tatsuya Amano, Harriet Bartlett, Dave Chadwick, Adrian Collins, David Edwards, Rob Field, Philip Garnsworthy, Rhys Green, Pete Smith, Helen Waters, Andrew Whitmore, Donald M. Broom, Julian Chara, Tom Finch, Emma Garnett, Alfred Gathorne-Hardy, Juan Hernandez-Medrano, Mario Herrero, Fngyuan Hua, Agnieszka Latawiec, Tom Misselbrook, Ben Phalan, Benno Simmons, Taro Takahashi, James Vause, Erasmus Ermgassen, and Rowan Eisner. 2018. 'The Environmental Costs and Benefits of High-Yield Farming'. *Nature Sustainability* 1: 477–485. https://www.nature.com/articles/s41893-018-0138-5.

Bauer, Mary. 2008. 'Migrant Tomato Workers Face Chronic Abuses'. Testimony before the US Senate Committee on Health, Education, Labour and Pensions. Southern Poverty Law Center. https://www.splcenter.org/news/2008/04/15/migrant-tomato-workers-face-chronic-abuses.

Baym, Geoffrey. 2010. *From Cronkite to Colbert: The Evolution of Broadcast News*. Boulder, CO: Paradigm.

Bishop, Claire. 2004. 'Antagonism and Relational Aesthetics'. *October* 110: 51–79.

Bishop, Claire. 2009. 'The Social Turn: Collaboration and its Discontents'. In *Rediscovering Aesthetics: Transdisciplinary Voices from Art History, Philosophy, and Art Practice (Cultural Memory in the Present)*. Eds. Francis Hall, Julia Jansen and Tony O'Connor. Stanford: Stanford University Press. 238–255.

Bishop, Claire. 2012. *Artificial Hells: Participatory Art and the Politics of Spectatorship*. London and New York: Verso.

Blair, Elizabeth. 2014. *All Things Considered: Revisiting The Stories Told in Murrow's 'Harvest of Shame'*. [Radio programme] NPR. Transcript. https://www.npr.org/2014/11/28/367244290/revisiting-the-stories-told-in-murrows-harvest-of-shame.

Bowe, John. 2003. 'Nobodies Annals of Labor'. *The New Yorker* 79.9: 106–133.

Brandes, Stanley. 1998. 'Iconography in Mexico's Day of the Dead: Origins and Meanings'. *Ethnohistory* 45.2: 181–218.

Broyes-Gonzalez, Yolanda. 1994. *El Teatro Campesino: Theatre in the Chicano Movement*. Austin: University of Texas Press.

Broyes-Gonzalez, Yolanda. 2006. 'Re-constructing Collective Dynamics: El Teatro Campesino from a Twenty-first Perspective'. In *Restaging the Sixties: Radical Theaters and Their Legacies.* 1st ed. Eds. James M. Harding and Cindy Rosenthal. Ann Arbor: University of Michigan Press.

Bruzzi, Stella. 2006. *New Documentary.* 2nd ed. London and New York: Routledge.

Bullard, Gabe. 2015. 'How to Update *Harvest of Shame* for the 21st Century'. *National Geographic.* https://www.nationalgeographic.com/people-and-culture/food/the-plate/2015/11/24/how-to-update-harvest-of-shame-for-the-21st-century/.

Canny, Autumn L. 2005. 'Lost in a Loophole: The Fair Labor Standards Act's Exemption of Agricultural Workers from Overtime Compensation Protection'. *Drake Journal of Agricultural Law* 10: 356–386.

Carruth, Allison. 2013. *Global Appetites: American Power and the Literature of Food.* Los Angeles: University of California Press.

Dove, Rita. 1994. *The Darker Face of the Earth.* London: Oberon Books, Ltd.

Fahlenbrach, Martin Klimke, and Joachim Scharloth, eds. 2016. *Protest Cultures: A Companion.* New York and Oxford: Berghahn.

'Farmworker Justice'. US Department of Labor. 2019. http://www.farmworkerjustice.org/wp-content/uploads/2019/05/NAWS-Data-FactSheet-05-13-2019-final.pdf.

Galati, Frank. 1991. *The Grapes of Wrath.* London: Josef Weinberger Plays.

Galati, Frank. 2011. '*The Grapes of Wrath*: Adapting Steinbeck'. https://www.youtube.com/watch?v=UHxMXTfDOs8.

Garza, María Alicia C. 2004. 'High Crimes Against the Flesh: The Embodiment of Violent Otherness in Cherríe Moraga's *Heroes and Saints*'. *Letras Femeninas* 30.1: 26–39.

Godoy, Maria. 2017. 'Delores Huerta: The Civil Rights Icon Who Showed Farmworkers "Si Se Puede"'. [Radio Programme] NPR. https://www.npr.org/sections/thesalt/2017/09/17/551490281/dolores-huerta-the-civil-rights-icon-who-showed-farmworkers-si-se-puede.

Greenberg, Linda Margarita. 2009. 'Learning from the Dead: Wounds, Women and Activism in Cherríe Moraga's *Heroes and Saints*'. *MELUS: Multi-Ethnic Literature of the U.S.* 34.1: 163–184.

Grossman, Marc. 2002. 'Chavez, Steinbeck: The Ties that Bind'. *The Sacramento Bee.* https://ufw.org/chavez-steinbeck-ties-bind-ufw-spokesman-longtime-cesar-chavez-press-secretary-marc-grossman-2/.

Harding, James M., and Cindy Rosenthal, eds. 2006. *Restaging the Sixties: Radical Theaters and Their Legacies*. Ann Arbor: University of Michigan Press.

Harum, S.M. Rafael and Yelena Ogneva-Himmelberger. 2013. 'Distribution of Industrial Farms in the United States and Socioeconomic, Health, and Environmental Characteristics of Counties'. *Geography Journal* 2013.385893. https://doi.org/10.1155/2013/385893.

Harvest of Shame. 1960. [TV programme] CBS. https://www.youtube.com/watch?v=yJTVF_dya7E.

Harvey, Fiona, Andrew Wasley, Madlen Davies, and David Child. 18 July 2017. 'The Rise of Mega Farms: How the US Model of Intensive Farming is Invading the World'. *The Guardian*. https://www.theguardian.com/environment/2017/jul/18/rise-of-mega-farms-how-the-us-model-of-intensive-farming-is-invading-the-world.

Haspel, Tamer. 17 March 2017. 'Illegal Immigrants Help Fuel US Farms. Does Affordable Produce Depend on Them?' *Washington Post*. https://www.washingtonpost.com/lifestyle/food/in-an-immigration-crackdown-who-will-pick-our-produce/2017/03/17/cc1c6df4-0a5d-11e7-93dc-00f9bdd74ed1_story.html?utm_term=.dbd6b9a32c14.

Hooks, Bell, and Amalia Mesa-Bains. 2018. *Homegrown: Engaged Cultural Criticism*. New York and London: Routledge.

Huerta, Jorge A. 1982. *Chicano Theatre: Themes and Forms*. Ypsilanti, MI: Bilingual Press/Editorial Bilingüe.

Huerta, Jorge A. 2000. *Chicano Drama: Performance, Society and Myth*. Cambridge: Cambridge University Press.

Intensive Farming: Advantages and Disadvantages. n.d. http://www.everythingconnects.org/intensive-farming.html.

Jackson, Julie. 2005. 'Theatrical Space and Place in the Presentational Aesthetic of Director Frank Joseph Galati'. *Theatre Topics* 15.2: 131–148.

Jacobs, Elizabeth. 2012. 'The Ecologies of Protest in the Theatre of Aztlán'. *Comparative American Studies: An International Journal* 10.1: 95–107.

Johnston, Hank, ed. 2009. *Culture, Social Movements, and Protest*. Farnham, Surrey and Burlington, VT: Ashgate.

Johnson, Ian. 2017. 'Industrial Agriculture Is Driving the Sixth Mass Extinction of Life on Earth Says Leading Academic, Professor Raj Patel'. *The Independent*. https://www.independent.co.uk/environment/mass-extinction-life-on-earth-farming-industrial-agriculture-professor-raj-patel-a7914616.html.

Juris, Jeffrey S. 2015. 'Embodying Protest: Culture and Performance within Social Movements'. In *Anthropology, Theatre and Development: The Transformative Potential of Performance*. Eds. Alex Flynn and Jonas Tinius. Basingstoke: Palgrave Macmillan. 227–247.

Kafer, Alison. 2013. *Feminist, Queer, Crip*. Bloomington and Indianapolis: Indiana University Press.

Kester, Grant H. 2004. *Conversation Pieces: Community + Communication in Modern Art*. Los Angeles: University of California Press.
Kuppers, Petra. 2007. 'Outsides: Disability Culture Nature Poetry'. *Journal of Literary Disability* 1.1: 22–33.
Lymbery, Philip. 2018. *Dead Zone: Where the Wild Things Were*. London, Oxford and New York: Bloomsbury Publishing.
Marchi, Regina M. 2009. *The Migration and Transformation of a Cultural Phenomenon*. New Brunswick, NJ: Rutgers University Press.
Martin, Carol. 2013. *Dramaturgy of the Real*. Basingstoke: Palgrave Macmillan.
Marx, Leo. 1991. 'The American Ideology of Space'. In *Denatured Visons: Landscape and Culture in the Twentieth Century*. Eds. Stuart Wrede and William Howard Adams. New York: Museum of Modern Art. 62–78.
McGreal, Chris. 9 March 2019. 'How America's Food Giants Swallowed the Family Farms'. *The Guardian*. https://www.theguardian.com/environment/2019/mar/09/american-food-giants-swallow-the-family-farms-iowa.
Mitchell, Don. 1996. *The Lie of the Land: Migrant Workers and the California Landscape*. Minneapolis and London: University of Minnesota Press.
Moraga, Cherríe. 1994. *Heroes and Saints & Other Plays*. Albuquerque, NM: West End Press.
Moraga, Cherríe. 2002. *Watsonville: Some Place Not Here and Circle in the Dirt*. Albuquerque, NM: West End Press.
Paolucci, Paul and Margaret Richardson. 2006. 'Sociology of Humor and a Critical Dramaturgy'. *Symbolic Interaction* 29.3: 331–348.
Patel, Raj. 2008. *Stuffed and Starved: The Hidden Battle for the World Food System*. London: Portobello Books Ltd.
Pawel, Miriam. 2014. *The Crusades of Cesar Chavez: A Biography*. New York and London: Bloomsbury Press.
Rancière, Jacques. 2004. *The Politics of Aesthetics*. Trans. Gabriel Rockhill. London and New York: Continuum.
Rancière, Jacques. 2006. 'Problems and Transformations in Critical Art'. In *Participation*. Ed. Claire Bishop. London: Whitechapel. 83–93.
Rancière, Jacques. 2010. *Dissensus: On Politics and Aesthetics*. Trans. Steven Corcoran. London and New York: Continuum.
Ratcliffe, Thomas N., and Lori L. Hall. 2014. 'Practicing the Art of Dissent: Toward a Typology of Protest Activity in the United States'. *Humanity and Society* 38.3: 268–294.
Rich, Frank. 1990. 'New Era for *Grapes of Wrath*'. *The New York Times*. https://www.nytimes.com/1990/03/23.theater/review-theater-new-era-for-grapes-of-wrath.html.
Shiva, V. 2000. *Stolen Harvest: The Hijacking of the Global Food Supply*. London: Zed Books.

Sobchack, Vivian. 1999. 'Toward a Phenomenology of Nonfictional Film Experience'. In *Collecting Visible Evidence*. Eds. J. M. Gaines and M. Renov. Minneapolis, MN: University of Minnesota Press. 241–254.
Solis, Octavio. 2015. *Alicia's Miracle*. Unpublished Manuscript.
Southern Poverty Law Center. https://www.splcenter.org/.
Steinbeck, John. [1939] 1992. *Grapes of Wrath*. New York: Penguin Group.
Taylor, Diana. 1997. *Disappearing Acts: Spectacles of Gender and Nationalism in Argentina's 'Dirty War'*. Durham and London: Duke University Press.
Thomas, Arden Elizabeth. 2016. 'Poisoning the Mother/Land: An Ecofeminist Dramaturgy in José Rivera's *Marisol* and Cherríe Moraga's *Heroes and Saints*'. *Theatre History Studies* 35: 143–160.
Union of Concerned Scientists. 2008. 'The Hidden Costs of Industrial Agriculture'. https://www.ucsusa.org/our-work/food-agriculture/our-failing-food-system/industrial-agriculture#.W-LbRnr7TYI.
United Farm Workers of America. *The Official Page of the United Farmworkers of America*. https://ufw.org/.
United Farm Workers. 2017. 'The 1965–1970 Delano Grape Strike and Boycott'. https://ufw.org/1965-1970-delano-grape-strike-boycott/.
United Farm Workers. 1986. *The Wrath of Grapes*. [YouTube] https://www.youtube.com/watch?v=Wq48o4ftL4A. or https://www.youtube.com/watch?v=GUOJE0Iie0E (Part 1) and https://www.youtube.com/watch?v=CmMgT-hHiLc (Part 2).
Valdez, Luis. 1994. *Luis Valdez—Early Works: Actos, Bernabé and Pensamiento Serpentino*. 2nd Printing. Kindle Edition. Houston, TX: Arte Público Press.
Van der Zee, Bibi. 4 October 2017. 'Why Factory Farming is not just cruel but a threat to all life on the planet'. *The Guardian*. https://www.theguardian.com/environment/2017/oct/04/factory-farming-destructive-wasteful-cruel-says-philip-lymbery-farmageddon-author.
Wald, Sarah. 2011. 'Visible Farmers/Invisible Workers: Locating Immigrant Labor in Food Studies'. *Food, Culture and Society* 14.1: 567–586.
Weisskopf, Michael. 1988. 'Pesticides and Death amid Plenty'. *The Washington Post*, 1–7. https://www.washingtonpost.com/archive/politics/1988/08/30/pesticides-and-death-amid-plenty/57bdfbce-6fcd-4c3a-b732-8f680566b630/?utm_term=.f776cf04dd94.
Yarbro-Bejarano, Yvonne. 2001. *The Wounded Heart: Writing on Cherríe Moraga*. Austin: University of Texas Press.
Yeung, Bernice, Kendall Taggart and Andrew Donohue. 2014. 'California's Strawberry Industry is hooked on Dangerous Pesticides'. *Reveal*, from the Center for Investigative Journalism. https://www.revealnews.org/article/californias-strawberry-industry-is-hooked-on-dangerous-pesticides/.

CHAPTER 6

Transforming Migrant Farmscapes: Performative Protest Art of the Coalition of Immokalee Workers

'I Too Am Human!'

In the twenty-first century, a leading voice for food justice in the American industrialised farm landscape is the Coalition of Immokalee Workers (CIW), a Florida-based human rights organisation that has improved working and living conditions for migrant farmworkers on large-scale produce farms from Florida to New Jersey. Through its grassroots political activism, community organising and efficacious protest art, the CIW transforms these sites that farmworkers have experienced as landscapes of exploitation and abuse into landscapes of justice and safety. While much has been written about the human rights work of the Coalition of Immokalee Workers from historical and political perspectives,[1] very little scholarship exists on the CIW's performative protest art.[2] This chapter, devoted exclusively to the work of the CIW, demonstrates how its partnering of art and activism has contributed to the success of their human

[1] See, in particular, Estabrook (2018), Giagoni (2011), Marquis (2017), and Minkoff-Zern (2014) as well as CIW's extensive website, and the websites of the Student Farmworker Alliance and Alliance for Fair Food. See also 'One of the Great Human Rights Success Stories' (2014) and 'CBS Sunday Morning: Fair Food Program' (2015). Much of the history of CIW in this chapter is gleaned from these sources, but I relied primarily on Marquis (2017) and interviews with CIW staff. When the information is available in multiple sources, I have not cited them.

[2] See Gouge (2016), Haedicke (2020) and Klein (2018).

© The Author(s), under exclusive license to Springer Nature Switzerland AG 2021

S. C. Haedicke, *Performing Farmscapes*, Performing Landscapes,
https://doi.org/10.1007/978-3-030-82434-1_6

rights campaigns and, in turn, to food citizenship. Vivid flags, banners, drawings, puppets, skits (or *teatros*) and music appeared from the beginning in the CIW's marches, vigils, public fasts and mobile museums. The protest art depicts the CIW's central narrative that casts migrant farmworkers as dignified human beings worthy of respect and justice for the essential jobs they do in graphic, inventive and comprehensible images[3] that raise public awareness of and engagement with issues around deplorable working conditions and tarnish retailers' brands. The 'I too am human!' phrase that appears on placards carried by farmworkers in marches and boycott rallies combatively contradicts the view that the farm labourer is a commodity.[4] Gerardo Reyes, a CIW staff member and former farmworker, comments in *Food Chains*, Sanjay Rawal's 2014 documentary on the Coalition of Immokalee Workers, that 'sometimes we as farmworkers don't eat. … To live hungry while you are working—that's not a dignified way of living. We want to be able to establish change!' (*Food Chains* 2014). Working hand-in-hand, the direct action and the protest art have actually changed the agribusiness landscape on many farms by challenging the 'official' narrative promoted by the industry touting the successes of industrialised farms to increase yield and lower food costs, a narrative that hides the damage of these intensive farming practices to land and labour.

The CIW is a worker-led organisation where farmworkers sit at the negotiating table rather than an organisation led by concerned outsiders advocating for the worker. The farmworkers determine and lead the activist and artistic strategies to effect social change. 'We are all leaders' is the CIW's motto, and it briefly encapsulates a key aspect of their

[3] See 'Alliance for Fair Food' (2018) video that shows some of the ways CIW partners the arts with human rights activism, https://www.youtube.com/watch?v=SYfKFmGNghQ&t=178s. The urls of videos and photographs appear in the footnotes in this chapter.

[4] See a photograph of the 'I too am human!' placards used in the fast outside Publix headquarters in 2012 in 'Publix: Radicals in the Heartland 2021', https://ciw-online.org/blog/2021/02/publix-radicals-in-the-heartland/. This recent blog entry links Publix to the 6 January 2021 insurrection to overturn the presidential election. These placards are still used today.

successful model of Worker-driven Social Responsibility (WSR).[5] The Worker-driven Social Responsibility Network writes:

> The WSR paradigm is founded on the understanding that, in order to achieve meaningful and lasting improvements, human rights protections in corporate supply chains must be worker-driven, enforcement-focused, and based on legally binding commitments that assign responsibility for improving working conditions to the global corporations at the top of those supply chains. (Worker-driven Social Responsibility Network 2020: n.p.)

The challenges facing the establishment of this kind of informed worker-driven approach to social change were obvious to the CIW from the beginning. In December 1997, inspired by Mahatma Gandhi and Cesar Chavez, six farmworkers staged a month-long hunger strike to protest low wages. The intended audience for the fast was not the growers, but rather the general public and the press. The fasting workers asked only to meet with the Florida growers to discuss fair wages. When asked why he would not even sit down with the CIW farmworker representatives, one grower responded angrily: 'The tractor doesn't tell the farmer how to run his farm' (Estabrook 2018: 108; Marquis 2017: 35; 'Because a Tractor' 2008). The CIW's flair for effective protest art emerged in the farmworkers' response. At the next demonstration, protesters wore white headbands with red lettering saying *Yo no soy tractor* [I am not a tractor] (Estabrook 2018: 108). The slogan summarises the core of the CIW's counter-narrative that farmworkers are human beings that deserve respect and justice. This conviction guides all their campaigns and often appears on the placards and other forms of art. A large mural of a tractor with the slogan and the signatures of farmworkers covering the image accompanied the workers on several campaigns. Now it hangs on the wall at the Coalition of Immokalee Workers' Community Centre in Immokalee. The CIW's model of Worker-driven Social Responsibility (WSR) is increasingly praised. In 2018, an article in *The Washington Post* by Gillian Thomas of the ACLU (American Civil Liberties Union) acclaimed the successes of the CIW: 'the results are stunning... given the exploitative starting point

[5] See Reyes (2018), '2019 in Review, Part 4' (31 December 2019), 'Worker-driven Social Responsibility—Beyond the Headlines' (2020), and WSR: Worker-driven Social Responsibility Network (2020).

of big agriculture, the transformation is nothing short of miraculous' (Thomas 2018).[6] In 2020, the Institute of Multi-Stakeholder Initiative Integrity (MSI Integrity)[7] completed 'Not Fit for Purpose: The Grand Experiment of Multi-Stakeholder Initiatives in Corporate Accountability, Human Rights and Global Governance',[8] a decade-long study of forty standard-setting programs to protect the human rights of workers in global supply chains. They found that the traditional model of corporate social responsibility (CSR), the major paradigm since the 1990s, was 'not fit for purpose' due to systemic flaws of relying on a top-down process instead of placing the worker in the leadership role and of failing to set up a system that holds corporations to account for forced labour and human rights abuses in their supply chains. The CIW's WSR model, it stated, has accomplished both those goals. In 2021, a study by researchers at University of Sheffield (UK), Stanford University (United States) and Yale University (United States),[9] confirming the inadequacy of CSR models, stated that '[t]ransparency legislation, a dominant mode of regulation, is not working' (Re: Structure Lab 2021: 5) and recommended replacing CSR models with 'binding worker-driven social responsibility agreements and meaningful worker empowerment' (Re: Structure Lab 2021: 24). The CIW has won numerous awards, notably the Franklin D. Roosevelt Freedom from Want Award (2013), the Global Citizen Award (2014), the Presidential Award for Extraordinary Efforts in Combatting Modern Day Slavery (2015), the ALBA/Puffin Award for Human Rights Activism (2018), and, in 2017, Greg Asbed, one of the co-founders of the CIW, won a McArthur Fellowship for his work in developing a 'worker-driven social responsibility' paradigm for labour justice ('CIW's Greg Asbed' 2017). Before looking at the important role CIW's art plays in its human rights activism, I will establish the context for its efficacy by providing a brief history of the organisation.

In the early 1990s, farms around Immokalee supplied 90% of the winter tomatoes bought in the United States, and Immokalee acted as the hub for thousands of migrant farmworkers who planted and

[6] The CIW website publishes many of the endorsements that appear in print media, news programs on radio and television, and documentaries.

[7] MSI Integrity, a non-profit organisation, grew out of the Harvard Law School's International Human Rights Clinic.

[8] See MSI Integrity (2020) and 'New Study Confirms' (2020).

[9] See Re: Structure Lab (2021).

harvested many of the fruit and vegetable crops up the east coast. In addition, Immokalee was often cited as having the worst working environment in US agriculture. What became the Coalition of Immokalee Workers began in 1993 with a small group of workers meeting in a room borrowed from a local church to discuss how to better their lives. Drawing on Paulo Freire's popular education model of understanding and analysing the situation, they began initiating or 'animating' reflective discussions about workers' rights and social change. Farmworker leaders, including Lucas and Ramiro Benitez, became active participants in these early discussions, as did local community specialist paralegals Greg Asbed and Laura Germino, who worked at the time for Florida Rural Legal Services (FLRS). FRLS worked on individual cases of worker abuse seeking damages for specific violations through the law, but Asbed and Germino wanted to do more. What if, they asked:

> instead of concerned outsiders advocating for the workers, the workers themselves came together as a community? What if instead of legal aid groups fighting individual legal battles, playing whack a mole with those caught violating the law, the workers' community fought for their human rights?...[Asbed and Germino] headed to Immokalee in hope of effecting cultural change in the direction of human rights and dignity, and not for a single victim at a time but for the entire community of workers. The two did not hide their purpose: 'We gave FRLS fair warning. We are coming down [to Florida] to work with the community on self-directed change'. (Marquis 2017: 11–12)

They believed that working with FRLS would offer them the opportunity to bring their combined skills (Germino's work on farmworkers' issues in the northeast United States and Asbed's involvement in community-led human rights action in Haiti) to Immokalee to effect social change for the whole community. Susan L. Marquis explains their approach:

> Greg and Laura argued that the traditional approach to advocacy did not take on the agricultural labor situation as a whole and therefore could not transform the system. This is where the most powerful element the two brought to Immokalee comes in. The changing of just a few words, from 'for' to 'with' and from 'individual' to 'community', belied what was in fact a radical change in perspective: instead of advocating *for* individual migrant workers, FRLS and other legal and social advocacy groups needed to be working *with* the worker community as a whole, 'finding a way for

> the [worker] community to analyse and understand its own situation so that it could bring its voice to the table with the growers'. (Marquis 2017: 16–17)

They believed that grassroots organizing strategies where communities organise themselves to effect worker-driven change could be effective with the tomato pickers in Immokalee, in part, because many of the newly arrived farmworkers had experience with worker-driven human rights organising in their homelands of Haiti, Guatemala, El Salvador and Chiapas (Marquis 2017: 17). Joining forces with other activists experienced in grassroots organising, like Cristal Pierre, Jean-Claude Jean, Pedro Lopez and Filipe Miguel, they formed the Southwest Florida Farmworker Project in 1993. First steps of the fledgling organisation included creating 'The Green Book', an illustrated pamphlet for the Immokalee farmworker community on their legal rights and conducting a participatory community survey, 'a month-long "listening" exercise…to learn what people in the community felt they needed to be changed' (Marquis 2017: 19). They also began offering Wednesday evening entertainment (films, music, skits/*teatros*) as a way to initiate or 'animate' reflective discussions about workers' rights and possible ways to combat injustices. These Wednesday evening dialogues pairing art and activism continue today. Led by farmworkers themselves, the discussions enable the community to work 'together to understand their situation and take responsibility for developing actions' (Marquis 2017: 24) to alter it.[10] This early activist work of the Farmworker Project made the focus on worker-driven social responsibility clear from the beginning. It revolved around an idea that was rooted in peasant organisations throughout Latin America and the Caribbean: Consciousness + Commitment = Change. This formula identified the three tools that still make up the core of their approach to human rights activism: consciousness through *popular education* and commitment through *leadership training* leading to change through *direct action and protest art*. These tools (discussed in more detail later in the chapter) grew out of Paulo Freire's radical pedagogy that sought to thwart the increasing dehumanization in modern society and to liberate the oppressed, primarily in Brazil and Peru, by helping them

[10] See Freire (1994: 46–48), for a lively dramatized example of how he 'animated' a discussion about the equivalence of the different 'knowledges' understood by him and by the workers.

develop a 'critical consciousness', an awareness of socio-political forces and constructions that circumscribed their lives. Liberation, Freire argues, is not just an abstract concept, but a practice that relies on reflection, dialogue and action. 'Critical consciousness', what Freire calls *conscientização*, means understanding how one's social reality is constructed and how it can be altered, and this critical awareness can lead to political action to effect social change (Freire 1994: 98–101).

The Farmworker Project's grassroots organising was receiving attention in the community and, in 1995, they called the first general strike by Immokalee farmworkers. Growers had applied pay cuts at the start of the tomato season and just a couple months later, Pacific Land Co., one of the largest growers, reduced wages further, down to the level of the 1970s. For five days, no crops were picked until Pacific Land Co. reversed the pay cut. What the workers understood from this direct action was that their community solidarity worked. Looking back twenty-five years later, a CIW blog entry reaffirmed that emphasis on solidarity: 'only when we realize that we are one community of workers with common interests, regardless of race or nationality or gender, can we truly come together to defend those interests as one, unified force' ('Yon Sel Ko' 2020: n.p.).[11] One of the farmworkers arrested during the strike was Haitian artist, Denis Remy. When he was released from jail, he created a painting honouring the coming together of multiple ethnicities of farmworkers in direct action. It depicts several ethnically diverse farmworkers raising their arms in the air to create one large fist surrounded by the words 'one strength' in Creole and Spanish. At the top and bottom of the painting is written 'The Coalition of Immokalee Workers' in English, Spanish and Creole.[12] The name stuck and the Southwest Florida Farmworker Project became the Coalition of Immokalee Workers (Fig. 6.1).

Another early event reinforced the power of community solidarity. In late 1996, Edgar, a novice tomato picker only sixteen at the time, stopped to take a sip of water as he harvested tomatoes in the hot sun. The crew leader attacked him, shouting 'You are paid to pick tomatoes, not drink

[11] See 'Yon Sel Ko' (2020) for a description of the Haitian peasant movement, Mouvman Peyizan Papay, that gave rise to this message of solidarity.

[12] See also 'Yon Sel Ko' (2020), https://ciw-online.org/blog/2020/03/yon-sel-ko/ and Zwerdling (2005) for a photograph of the painting, now hanging in CIW's community centre.

Fig. 6.1 Denis Remy's painting on the wall of the CIW community centre. Photographer: Susan Haedicke

water!' (Marquis 2017: 27).[13] Edgar, bloodied and weak, staggered back to Immokalee to the CIW offices. A few days later after long discussions about how to respond, several hundred farmworkers marched to the crew leader's house. They carried Edgar's bloodied shirt as an emblem and chanted 'golpear a uno es golpear a todos!' (when you beat one of us, you beat us all!). While a line of police created a wall between the farmworkers and the crew leader, one of the protesters and co-founder of the CIW, Lucas Benitez, realised that the police were on their side as he saw some officers giving a surreptitious 'thumbs up'. The CIW had taken 'their first steps in the direction of systemic change' (Marquis 2017: 29) in the agricultural landscape as a whole, and it confirmed the strength of coming together. That bloody shirt was not washed, and it was not thrown away. It has become an icon in the CIW's symbology, and it is occasionally

[13] One of the popular education paintings depicts such an event. See 'Fair Food Program Education: Travels to Georgia, South Carolina, North Carolina' (2017), https://ciw-online.org/blog/2017/06/photos-ffp-education/.

displayed to make a point. In 2011, for example, it was carried on a bicycle march from Immokalee to Naples for a ribbon-cutting ceremony for a new grocery store in the Trader Joe's chain to highlight the importance of joining the Fair Food Program. A CIW member explains, 'Trader Joe's is famous for employees' quirky Hawaiian shirts; it's a key component of their brand identity. In Immokalee, we also have an iconic shirt that embodies what we are about'.[14] Trader Joe's signed onto the Fair Food Program soon after. The early successes in battling injustice in the Florida tomato industry helped define their organising strategies for the next several years and led to the launch of The Campaign for Dignity, Dialogue and a Fair Wage. The headbands printed with 'Yo no so tractor', the painting giving the CIW its name and the bloody shirt introduce the role that art plays in the CIW's human rights activism. Clearly, the farmworkers in CIW had a flair for protest art that would draw attention to the issues and clarify their message. It is this intersection of politics and aesthetics, evident in these early examples and developed more fully into a wide range of art forms in subsequent years, that has contributed to their successful campaigns that have altered the migrant farmscape in the eastern United States.

Activist artist, David Solnit, who has worked closely with the farmworkers in Immokalee to design the art in some of their public campaigns, rallies and pageants for close to twenty years, affirms the power of the arts: 'skills involved in the arts and performance are essential if there's going to be radical change on the planet' (Solnit 2014a).[15] For Solnit, core conflicts in society arise from fights between opposing narratives, a 'battle of stories' (Angel 2008: n.p.; Solnit 2012) told in words, images and performances, so the need for compelling narratives is key to political movements that seek radical social change. Sociologist Francesca Polletta concurs: 'stories are better able than other kinds of messages to change people's opinions' (Polletta and Chen 2012: 487) because 'we immerse ourselves in the story, striving to experience vicariously the events and emotions of the protagonist's experience [and] this experience of immersion or "transportation" can lead to lasting changes of opinion' (Polletta 2008: 27). She argues that we as readers 'expect to have to

[14] 'Farmworkers to Bike from Immokalee' (10 February 2011). See also Marquis (2017: 27–28). See also 'Not 1996 Anymore' (15 March 2012).

[15] See also Solnit (2014b and 2016).

interpret stories' (Polletta 2006: viii), to fill in the gaps, and that creative act strengthens audience engagement.[16] Echoing Paulo Freire, Solnit claims that the 'system' keeps people in their place through narratives that reinforce the status quo, presenting it as normative and unchangeable.[17] The way to resist is to create counter-narratives that deconstruct the opponent's story and tell compelling new stories. Philosopher Hilde Lindemann Nelson argues that counter-narratives (or 'counterstories' as she calls them) undo socially shared narratives in which certain communities' identities have been defined by those in power rather than by the community itself. Counterstories 'repair' those 'damaged identities' (Nelson 2001) narratively as individuals place their own stories of dignity and equality in contrast to the 'official' narrative. The counterstories provide the basis for social action.[18] Solnit has worked closely with the CIW to create visual and performance-based art for worker-led counter-narratives. Artists, he insists, are particularly adept at telling stories that communicate the 'heart and gut and not just facts and figures' (Solnit 2012).[19] They do not just decorate the protest; they shape direct action so that it is clear and powerful. Since artists excel in story-telling skills, they need to be key members of the team, and their art-making strategies must sit side-by-side with traditional organising strategies (Solnit 2014a, b).[20] Sociologist Melissa A. Gouge applies these ideas to the work of the CIW: 'The CIW uses subversive storytelling and creative playfulness deliberately to build a common worker identity. They tell an alternative history

[16] See also Polletta (2009, 2015), Polletta et. al. (2011), and Polletta and Callahan (2017).

[17] In his 'spoken book' with Antonio Faundez, Freire explains the omnipresence of resistance: 'Sometimes in our uncritical understanding of the nature of the struggle, we can be led to believe that the everyday life of the people is a mere reproduction of the dominant ideology. But it is not. There will always be something of the dominant ideology in the cultural expressions of the people, but there is also in contradiction to it the signs of resistance—in the language, in music, in food preferences, in popular religion, in their understanding of the world' (Freire 1989b: 26).

[18] See also Nelson (1995) and Dixon (2018).

[19] See also Renz (2005).

[20] CIW employs visual artist Marley Monacello on permanent staff. All the art is created with or by the farmworkers.

through radical media that creates a space for transnational[21] solidarity-building among farmworkers and participants in the movement' (Gouge 2016: 862). She argues that the 'self-conscious use of creative playfulness and subversive storytelling...shapes the relationship between human rights and labor rights' (Gouge 2016: 862). Although always visible in the visual documentation of the activism of the Coalition of Immokalee Workers, the artwork is rarely researched in spite of the fact that it acts as one of the cornerstones of CIW's human rights activism,[22] and it has contributed to altering not only public perception of industrial farm landscapes, but the actual lived landscapes for farmworkers and growers as well.

Like the case studies discussed in the previous chapter, the CIW's protest art often relies on techniques of collage, as defined by Jacques Rancière (discussed in Chapter 5), to produce a shock that can propel people who were unaware of the situation into action, but the Coalition goes one step further as their activism and art also offer solutions. While Grant H. Kester recognises the power of 'shock' to propel spectators into awareness and critical reflection (Kester 2004), he questions how well it can initiate insight and increased agency and claims that the avant-garde's use of aesthetic shock is about leading audience responses (Kester 2011: 183). The disorientation caused by shock, Kester claims, resembles trauma in its psychological impact and thus arrests new insights and locks the viewer in 'continual re-enactment, repetition, or reiteration of the traumatic event' (Kester 2011: 183). Rather, he advocates an alternative model of dialogic collaboration that 'occasions a reciprocal testing of both ethical and aesthetic norms, the outcome of which can only be determined through the subsequent forms of social interaction mobilized by a given work' (Kester 2011: 185). The performative protest art that accompanies the CIW's human rights campaigns, I think, often manages to successfully combine aesthetic shock and dialogic collaboration. The

[21] Gouge says the 'I use the term "transnational" here to refer also to practices, strategies and identities that seek to cross boundaries distinguishing immigrant labor from allied communities, and farmworkers from different indigenous communities throughout the Americas' (Gouge 2016: 862).

[22] Much of the art is available online on both the Coalition of Immokalee Workers website and in videos documenting their work. See *CNN Freedom Project* (2017a), a series of three videos on various campaigns. I have received additional information from the interviews and informal discussions I have had with CIW staff: Greg Asbed, Marley Monacello, Julia Perkins, and Nely Rodriguez.

aesthetic shock comes from the compelling stories of the working conditions in the Florida tomato fields performed in a range of art forms in the public demonstrations, but that shock is tempered by the dialogic collaboration that happens not only between the farmworkers and the public, but also among farmworkers in the worker-led popular education sessions and art-making for the performative protest art that accompanies the direct action. While much of the CIW's dialogic collaboration occurs in the form of discussion (or 'reflection'), a more participatory and performative version is evident in the 'yarn web' where the *discussion* appears as a temporary sculptural web, often on the street as part of a protest. Participants of the protest and sometimes passers-by create a circle. One participant holds onto the end of the yarn rolled into a ball, starts a discussion with a thought and a toss of the ball of yarn to another participant who catches it, contributes another thought to the discussion and throws it on to yet another participant. Each participant holds onto the string at the place where he or she caught it. Comment after comment and string after string are added to the circle, and the resulting web embodies the complicated conversation.[23] This dialogic collaboration shifts the discussion from words to shared actions creating impromptu art. This yarn web also displays the CIW's use of what Gouge calls 'creative playfulness': 'an action, a mood, and/or an orientation that…is joyful, agentic, communicative and improvisational. It is also interactional, subversive to social hierarchies, and pre-figurative of democratic politics' (Gouge 2016: 864). Playfulness and humour are important to political action as they can highlight inconsistencies in official narratives.[24]

For Solnit, art can show the way to the future because it goes beyond desire for change to celebration of hope and determination to win the 'battle of stories' (Solnit 2016). Solnit's words again echo Freire as he seeks to understand 'the process in and by which things come about' (Freire 1994: 16). Freire argues that:

> the future of which we dream is not inexorable. We have to make it, produce it, else it will not come in the form that we would more or less wish it to. True, of course, we have to make it not arbitrarily, but with

[23] For a photograph of this technique used in the Freedom Fast in New York City in 2018, see 'Effects of Fast' (2018), https://ciw-online.org/blog/2018/03/freedom-fast-day-2/. Scroll down about half way.

[24] See Paolucci and Richardson (2006).

> the materials, with the concrete reality, of which we dispose, and more as a project, a *dream*, for which we struggle. (Freire 1994: 101)

Aesthetic shock, dialogic collaboration and 'creative playfulness' together provide tools to make Freire's 'project, a *dream*' into a reality as the CIW counter-narratives expose the horrors in the agri-business landscape, but simultaneously insist that the farmworkers are not helpless victims, but rather strong partners with the growers and retailers all the way up the food chain in their endeavour to counteract the degradation and exploitation of labour and land. Gerardo Reyes explains:

> We [workers] can be portrayed in two lights. One is as 'victims' and our stories can be compelling. People then act because they pity us. Or, we can be portrayed as architects of the future we want. Then the work we do is the work to protect our dignity as workers. It is difficult to *inspire* others if you are put in a box of 'victim'. (Marquis 2017: 58)

Transforming Migrant Farmscapes: CIW's Campaigns

In the 1990s, the CIW followed tactics of strikes and protests used by labour unions, particularly the United Farm Workers Union (UFW), although with significant differences: the CIW represented all farmworkers in Florida's tomato industry, not just union members, and the changes it sought were across the industry, not just on particular farms. In addition, the agricultural landscape had changed in the years since the UFW began its labour organising. Through the CIW's ongoing analyses of the situation that keeps farmworkers in poverty, they gradually realized that today's villains are not the individual growers satirised in Valdez's *actos* (discussed in Chapter 5), but rather the fast-food corporations and large grocery store chains that 'negotiate' (or force) reduced prices from their suppliers (the growers) in the interest of low-cost food. These reductions have an impact on the growers' income and, going down the food chain, on farmworkers' wages.[25] The Coalition's understanding of that changed landscape began with the knowledge from several sources about the inability, rather than a stubbornness, of the growers to pay higher

[25] The 2014 documentary, *Food Chains*, makes the retailers' complicity in low wages, poor working conditions, and violence in the fields blatant in its animated graphics.

wages (Marquis 2017: 48). Then in 2001, at one of their evening meetings, the farmworkers were discussing an article published in *Packer* in which Taco Bell had bragged about 'negotiating' a lower price for tomatoes from their suppliers. One worker asked if Taco Bell could force prices down, could the Coalition pressure Taco Bell to force prices up. That 'light bulb moment' helped change their 'perspective and strategy' (Marquis 2017: 52). As they looked up the food chain, they could see that their poverty wages could only be addressed in dialogue with the corporate buyers, and that dialogue could be imposed on the large fast food and grocery store chains since they had a brand to protect.[26] Consumers return over and over to a familiar and reliable brand but, as the CIW realised, the staying power of the brand was also a source of vulnerability if the brand became tarnished and so drove customers away. What if, they asked, the fast-foods chains' 'public performance' enacted through its brand could be reframed from familiar and reliable to exploitative and abusive? What if the restaurant-goers had to decide whether they could eat food harvested by modern-day slaves? What if the CIW could link the tomato on the plate to sexual abuse in the field? The Coalition saw an opening to partner with customers to force the fast-food corporations to take responsibility for justice in their supply chains. This new strategy led to the Campaign for Fair Food in 2001 and the subsequent Fair Food Program in 2011 that developed a Code of Conduct ensuring fair wages and safe working conditions. The CIW's Campaign for Fair Food had three key demands of corporate retailers: (1) one penny more for each pound of tomatoes picked, paid by participating corporate retailers and passed onto the workers in the form of a bonus, (2) a commitment by retailers to buying tomatoes only from growers who abide by the Fair Food Code of Conduct and (3) a direct partnership with the farmworkers to design, implement and enforce the fair food demands. After ten years of creative, tireless campaigning, the CIW's agreements with major food retailers allowed for the creation of the Fair Food Program in 2011, and its Fair Food Code of Conduct went live in the fields, providing unprecedented, verifiable and enforceable human rights standards in the fields, including zero tolerance for forced labour; zero tolerance for physical and verbal abuse as well as sexual assault; access to clean drinking water, shade and toilet facilities in the field; and the right to report abuse without fear

[26] See Marquis (2017: 49–51), McMillan (2015/2016) and 'Modern Farmer Article' (2015) on the story of that realization.

of retaliation. The Fair Food Standards Council (FFSC), the CIW's independent partner organisation, rigorously monitors the participating farms through frequent audits and a 24/7 complaint hotline. The Fair Food Program was the first fully-fledged application of the Worker-driven Social Responsibility Model, establishing a worker-led approach supported by both popular education and leadership training programmes for farmworkers and real market consequences when workers' rights are violated. At the time of this writing, fourteen multi-billion-dollar retailers, both fast food and supermarket chains, have joined the Fair Food Program to end exploitation, abuse and poverty wages. Taco Bell was the first to sign on.[27]

Another contributing factor to the CIW's success that is less frequently discussed is the on-going role that the counter-narratives of the protest art play in the campaigns. This visual and performative storytelling translates incoherent moments of the struggle into a comprehensible language that can engage the general public and politicians and make a memorable negative impact on a corporate brand.[28] As film theorist Stella Bruzzi explains, 'the raw material is incapable of drawing out or articulating the truths, motives, or underlying causes it both contains and implies, so it falls to the writer to extract this general framework' (Bruzzi 2006: 13). For the Coalition of Immokalee Workers, while all the factual information and the testimonies of exploitation and abuse in the fields have dramatic potential, that story needs a storyteller to sift through and craft all the details. The CIW's performative art has honed the extensive material documenting farmworker exploitation into compelling politico-dramatic narratives that portray migrant farmworkers as essential and dignified workers, not as victims or commodities. What makes these narratives so compelling however is their staging of the events through a recognisable 'David and Goliath' tale transplanted into contemporary realities that underlie each of the many art forms. The battle between the farmworkers and the retail giants is blatantly uneven, but as David, represented by

[27] While the details of the boycott of Taco Bell, the alliances established and other tactics used in the campaign that convinced Taco Bell to agree to the tenets of the Campaign for Fair Food in 2005 are outside the scope of this book, Marquis writes a detailed history of the Taco Bell campaign (Marquis 2017: 52–77). Her book provides extensive information on later campaigns as well.

[28] For several photographs of the marches and vigils in the CIW's Campaign for Fair Food, see J. J. Tiziou (2017), http://www.jjtiziou.net/series/ciw-fair-food.

the farmworkers, is joined by the public, he grows more powerful than Goliath so guaranteeing a successful outcome. This familiar narrative has helped establish strong alliances between the farmworkers, students and religious leaders and built a sense of solidarity between the public and the farmworkers to hold food giants accountable for human rights abuses in their food chains. It also illustrates the power of coalition-building, one of the key dimensions of food democracy (Hassanein 2008).

Taco Bell, managed by Yum! Brands, was the first retailer that the CIW chose to challenge, and the campaign with marches and boycott rallies showed their propensity to use efficacious performative art as a part of their activism. One narrative placard told the public: 'Yes, there are REAL tomatoes in Taco Bell food and REAL people pick them who haven't had a raise in twenty years!' Others shaped like hands asked questions about who picked the food, and tomato-shaped posters grin maliciously at the public. An over-sized two-dimensional puppet of Taco Bell's mascot—a chihuahua—joined the marches and participated in agit-prop skits (teatros) that clarified the issues of the campaign.[29] In one early teatro during the boycott of Taco Bell, the chihuahua puppet, representing Taco Bell, proposes to Queen Cheap Tomato (a large grinning red tomato puppet representing the tomato industry). The festive wedding ceremony begins, but when the priest asks if anyone can show just cause why this couple should not be joined in matrimony, the celebratory event begins to fall apart as opposition from a farmworker, a student and a religious leader expose 'just cause' as they each show how this 'union' leads to injustices in the fields. The wedding is stopped, and the event transforms into a celebration with singing and dancing as the marchers and onlookers join the farmworker-performers. Soon after it was performed in front of Taco Bell's headquarters, CIW farmworkers sat at the negotiating table with Taco Bell's directors, and Taco Bell became the first multi-billion-dollar fast-food chain to sign onto the Fair Food Program in 2005. It was followed by McDonald's in 2007, Burger King and Subway in 2008, Chipotle in 2012, and Yum! Brands extended its Fair Food agreement to all its chains, including Pizza Hut and Kentucky Fried Chicken. Large grocery store chains, including Whole Foods, Trader Joe's and Walmart and more, were not far behind. And the CIW has shared its

[29] See Bacon (2002) for an early photograph, https://inthesetimes.com/article/no-quiero-taco-bell. See also 'Coalition of Immokalee Workers: La Mixteca' (2003).

model of Worker-driven Social Responsibility with other industries across the globe.[30]

Migrant Farmscapes and Popular Education

Enabling farmworkers to sit at the negotiating table relies on Freirean popular education models to ensure that workers understand their rights and on leadership training to help them use their voices. Freire called popular education a 'pedagogy of hope' that enabled 'the popular classes to develop their language: not the authoritarian, sectarian gobbledygook of "educators," but their own language—which, emerging from and returning upon their reality, sketches out the conjectures, the designs, the anticipations of their new world…language as a route to the invention of citizenship' (Freire 1994: 39). And, indeed, a mural above the windows in the CIW Community Centre shows Chavannes Jean Baptiste, Co-founder of the Haitian Peasant Movement of Papaye and long-time practitioner of Freirean Popular Education methods, animating a discussion with co-conversants. Art plays a large role in the CIW's popular education since visual and oral modes of communication are more effective than the written word in explaining their rights to farmworkers because, as Greg Asbed explains, the community is 'made up of people who are putting most of their energy into survival and who have limited formal education' (Marquis 2017: 22). Cartoons, songs and *teatros* vividly depict the reality of the Florida tomato field landscape. One early cartoon depicted a farmworker in the tomato field lifting the familiar tomato bucket up to the grower who tosses him a coin. A price tag saying '$7,500 por año' points to the small pile of coins at the farmworker's feet. The grower representing the tomato farm industry sits on a level above the worker next to a larger pile of dollar bills identified with a price tag saying '$120,000,000 por año'. He hands tomatoes to the chihuahua mascot for Taco Bell[31] who, seated on yet a higher level next to a pile of money

[30] See 'Climate Underground' (2020). See also '"Future of Work" Round Table' (2018), Greenhouse (2018), MSI Integrity (2020), 'New Study Confirms' (2020), and 'Worker-driven Social Responsibility—Beyond the Headlines' (2020). And in May 2021, Anita Hill's Hollywood Commission established a partnership with CIW to adapt WSR models to the entertainment industry. See 'Anita Hill-Led Hollywood Commission' (2021).

[31] See Marquis (2017: 50).

labelled '$5,200,000,000 por año', tosses the grower a dollar bill. With his other hand the mascot hands food to the consumer eating a taco as he sits on the highest level next to a huge pile of money—'all the money in the world'. He tosses two bills to Taco Bell as he says 'Thanks for the chalupa, Dude!' (Fig. 6.2). This cartoon echoes El Teatro Campesino's 1966 *acto*, *Quinta Temporada*, where the farmworker harvests dollar bills hanging from the character, Summer, as fast as he can and stuffs the money into his back pockets. The crew leader, Don Coyote, stands behind him taking the money out of the worker's pockets just as fast and handing it over to the grower, Patron, who stands behind him (Valdez 1994: loc. 432).

The Coalition also uses the *teatro*, similar to El Teatro Campesino's *acto*, in their Wednesday night community-building sessions for workers. Each *teatro* is devised and performed by veteran farmworkers to address commonly held beliefs, especially among new arrivals. The *teatro* has

Fig. 6.2 Early Cartoon. 2002. Photograph by the Coalition of Immokalee Workers

proven to be a successful way to encourage dialogue and change attitudes and behaviour. The Coalition's *teatro*, *Door of Fear*,[32] satirically exposes the devastating consequences of giving in to paralysis brought on by fear—of deportation, exploitation, abuse, loneliness and loss of job—that many farmworkers feel. The *teatro* begins as the 'king' holds court over farmworkers who have committed petty crimes. He offers one worker after another a choice between execution and a step into the unknown by going through a curtain that hides what is on the other side. Each of the farmworkers, paralysed by fear, accepts death. After several farmworkers have been shot with rubber tipped arrows and fall clumsily to the floor, the executioner asks what is so dreadful behind the curtain that all would choose death. The king tells her to find out for herself. Although she trembles and resists, the king encourages her and when she finally gets up the courage to peek, she discovers a sunny and bountiful world on the other side—a world of social justice without fear that paralyses the farmworkers into inaction. When she asks the king why he scared them, he responds 'I did not scare them. I was giving them an opportunity....I gave them all the option of going there'. He does not tell the workers what is on the other side because the farmworkers must discover that for themselves for the 'new world' to be successfully theirs. The farmworkers must know they are capable of learning and acting on their own.[33] In the discussion (or 'animation') afterwards, a CIW staff animator enters into dialogue with the audience of farmworkers not to 'deposit' information into their heads as in what Freire calls the 'banking concept of education' (Freire 1989a: 58), but rather to initiate an inquiry or 'problem-posing education [that] rejects communiqués and embodies communication' (Freire 1989a: 66). Liberation is achieved through education that depends on dialogic relations:

> Through dialogue, the teacher-of-the-students and the students-of-the-teacher cease to exist and a new term emerges: teacher-student with students-teachers. The teacher is no longer the-one-who-teaches, but one who is himself taught in dialogue with the students, who in turn while being taught also teach. They become jointly responsible for a process in which all grow! (Freire 1989a: 76)

[32] For a video clip, see https://www.youtube.com/watch?v=77nWemjYUbY&t=4s and 'Off Broadway...in Immokalee!' (2018).

[33] See also Freire (1994: 44–49).

For Freire, '[k]nowledge emerges only through invention and re-invention, through the restless, impatient, continuing, hopeful inquiry men pursue in the world, with the world, and with each other' (Freire 1989a: 58) and '[a]uthentic liberation—the process of humanization is not another deposit to be made.... Liberation is a praxis; the action and reflection of men upon their world in order to transform it' (Freire 1989a: 66). After the *Door of Fear teatro*, the CIW animator, following the example of Freire's animator (Freire 1994: 46–48), asks the audience of farmworkers to think about times when they 'hesitated to do something out of fear. Because you did not know what would happen? And how many times have you overcome that fear, and taken a step into the unknown?... We also have to ask: who wins when we are afraid?' ('Off Broadway...in Immokalee' 2018: n.p.). The audience of farmworkers shouts out answers: 'The bosses, the supervisors. Corporations, like Wendy's' ('Off Broadway...in Immokalee' 2018: n.p.). As farmworkers share stories of fear from the fields, they also speak of moments of solidarity and resistance that garner results. In words that sound like Freire, the animator continues:

> The people who cross over a frightening threshold do so because they have hope. Their hope overcomes their fear. They are carrying a dream. So they go forward, even when the door may appear frightening. .... When we first began the Campaign for Fair Food, that was scary. We were afraid. But we dared to cross that threshold anyway, we overcame our fear, carrying our hope. Today, after two decades, no matter who has been in power in Washington, we have won real change for our community — something that for many generations of farmworkers was unattainable. ('Off Broadway...in Immokalee' 2018: n.p.)

One worker's response revealed the Freirian efficacy of the process: 'At the beginning, you unlearn what you know about the world and start seeing it from a different perspective. You start examining the system' (Marquis 2017: 22).[34]

The *teatro* is also used to create alliances, particularly between farmworkers and students at the annual Student/Farmworker Alliance Encuentro—a gathering of students from across the United States in

[34] For a summary of another *teatro* on the importance of solidarity, see Haedicke (2020).

Immokalee to learn about and participate in the CIW's activism. At the three-day long September Encuentro 2018,[35] the discussions focused on developing a strategy to involve students in the on-going Wendy's boycott. Wendy's is the sole American fast-food giant to refuse to join the Fair Food Program. In 2018, the chain announced that it would no longer buy tomatoes from Mexico because of the well-documented abuses of farmworkers and instead would purchase American and Canadian tomatoes grown in greenhouses, and it established its own supplier Code of Conduct, in part, as a response to the CIW's five-day public fast and protest rally in front of Wendy's headquarters in New York City in March and again in July 2018 (Lee 2018: n.p.). Buying tomatoes grown in greenhouses was not a viable solution for the CIW since working conditions there are very bad and sexual abuse is rampant, and the CSR code of conduct model is inadequate. *At the Encuentro*, the CIW used a *teatro* to expose Wendy's complicity in farmworker abuse. The *teatro* opens as farmworkers carry the 'greenhouse', a metal structure covered in plastic on stage. They quickly fill the greenhouse with cardboard plants and the recognisable red tomato buckets and affix a sign saying Don Fresco's Produce. The farmworkers enter the greenhouse to begin picking tomatoes. Conditions of extreme heat, lack of access to drinking water and exposure to pesticides as a supervisor sprays a canister with a skull-and-crossbones into the greenhouse with the workers inside begin to take their toll. When one worker asks for water, the boss sends her back to work and tells her to stop complaining. And when another worker protests that her co-worker is sick from the spray, the boss ignores her concerns. The familiar 'Wendy's' with her freckles, red hair braided and tied with blue bows and blue and white striped dress appears. The boss quickly orders the worker back to work to tell all the workers to look happy as he rushes to set out a water jug. 'Wendy's' and the boss mutually praise the great working conditions and contentedness of the workers, even though both know the opposite is true. When one worker in the greenhouse faints, the boss ostensibly distracts 'Wendy's' attention from what is happening with a gift of tomatoes. They proclaim success in the decision to move the tomatoes to greenhouses as the taste of the tomatoes is great, the working conditions are safe, and the workers are happy. As soon as 'Wendy's' leaves, the boss harshly reprimands the workers for not working hard

[35] See 'Scores of Students' (2018) and 'Over 80 Students Gather in Immokalee' (2018) for extensive details and photographs of the Encuentro 2018.

Fig. 6.3 Teatro about Wendy's treatment of farmworkers for the Encuentro 2018. Photograph by the Coalition of Immokalee Workers

enough, yells at the worker who fainted, and orders the other workers to take her away (Fig. 6.3).[36]

As with the other *teatros,* animators and students reflect on what they have seen and relate it to what is happening in the Campaign. This *teatro* was followed by campaign strategy and skills-building workshops leading to an actual boycott picket on Day 3 (22 September) of the Encuentro 2018 at a Wendy's restaurant in Estero, Florida. The protest gathered a crowd of over 150 people, and several participants tried to deliver a letter with calls for justice ('Over 80 Students' 2018; 'Scores of Students' 2018). The Encuentro 2018 led to the Student-Farmworkers Alliance 'Pulling Back the Curtain on Wendy's' week-long campaign on

[36] Information about this *teatro* came from my interview with Nely Rodriguez (5 October 2018) and an email from Julia Perkins (21 February 2020).

university campuses across the country in October 2018.[37] It sought to put pressure on university administrators to expel Wendy's from their campuses as part of the on-going 'Boot the Braids' campaign,[38] started in 2014. The Spring Action 2019 '4 for Fair Food Tour' ('Major Action Announced' 2018) again focused on eliminating Wendy's from university campuses. The protesters travelled to four flagship state universities—University of Michigan, Ohio State University, University of North Carolina-Chapel Hill and University of Florida—over a two-week period in March and joined forces with the students to create performative marches with flags, banners and puppets of the university president and Wendy's mascot.[39] During the Campaign, University of Michigan ended its contract with Wendy's, and student unions at other universities voted to ban the fast-food chain from their campuses. The Wendy's campaign is still on-going.[40]

Paintings specifically created for popular education sessions that take place in the Wednesday night community meetings and at training sessions in the fields create performative and participatory versions of the large-scale tomato farmscapes. A painting visually communicates a key issue or situation often experienced by the farmworkers and initiates an *animation* about what the image is depicting and what can be done to change the situation. As CIW staff member, animator and former farmworker Nely Rodriguez describes, this dialogic collaboration becomes both a problem-posing session exploring ways to change the real-life situation and a leadership training as the farmworkers engage in analysis of the situation depicted: 'critical consciousness' based on Freire's pedagogy (Personal interview, 5 October 2018). Since the painting concepts are created collaboratively by long-time farmworkers, Rodriguez says that the workers in the popular education sessions can immediately recognise the familiar situation and identify the problem the painting is addressing. For Freire, the starting point for knowledge acquisition and construction must

[37] See a video clip of a student-devised Wendy's agit-prop *teatro* at University of North Carolina, https://vimeo.com/298878467. See also 'Photo Report, Part 2' (2018) and 'Photo Report' (2018).

[38] The title 'Boot the Braids' refers to the Wendy's mascot of a young girl with her hair plaited into two braids. See also 'Action Round-Up' (2018).

[39] See 'A Truly Grand Finale!' (2019) and '4 for Fair Food' (2019).

[40] See CNN 'Freedom Project—Wendy's Boycott' (2017b). The CIW website continues to update the Wendy's campaign.

always be in the 'here and now' of the student or the journey to 'there' will never happen. 'You never get *there* by starting from *there*, you get *there* by starting from *here*' (Freire 1994: 58. Italics in original).

One of the drawings focuses on worker-to-worker sexual harassment (Personal interview, 5 October 2018) (Fig. 6.4). It depicts a smirking male farmworker carrying his bucket on his shoulder leaving his other hand free. In the narrow alley between the tomato rows, he walks past a female farmworker bending over to harvest tomatoes. As he passes her, he makes an offensive remark, '¡Ai Mamacita! ¡Qué rica te ves!' roughly translated to 'Hey there, baby! Looking real hot…'. In the popular education sessions that Rodriguez leads using this painting, she finds laughter to be the immediate male response. She asks them to read the comment and narrate the scenario of the painting. She then asks a question: 'if this were your mother, your wife, your daughter hearing a comment like this,

Fig. 6.4 Fair Food Program Sexual Harassment Popular Education Drawing. 2018. Ideas generated by farmworkers. Photograph by the Coalition of Immokalee Workers

would you like it?' The reaction is always 'stunning', she says, and the environment changes immediately as the laughter switches to loud shouts of 'No!' (Personal interview, 5 October 2018). Kester's dialogic collaboration begins here as the group listens to the female farmworkers describe how comments make them feel, explores the impact of feeling safe has on productivity and the lived landscape in the fields and clarifies why sexual harassment cannot be tolerated. The popular education session is often held on the farms participating in the Fair Food Program, so farmworkers, crew leaders and everyone else working in the fields attend and learn about workers' rights under the Fair Food Program[41] and why zero tolerance of sexual harassment is good not only for women pickers, but for all pickers.

Another new drawing for the 2018–2019 tomato season focuses on workers' rights to report abuse. As part of participating in the Fair Food, growers agree to permit Fair Food Standards Council (FFSC) investigators to visit the fields throughout the growing season and talk to farmworkers about the working conditions. While they have the right to talk to investigators, farmworkers occasionally say they still feel pressure from the supervisors to say nothing. One painting shows a field supervisor listening to the conversation between the farmworker and the investigator, and the farmworker looks 'scared' by the situation (Fig. 6.5). This painting leads to reflections that often start with the question of how many have been told to say nothing to investigators or not to call the 'Fair Food people'. Many raise their hands and they discuss what they can do. Its companion painting shows a 'dumper'[42] scolding a farmworker for speaking to a FFSC investigator (Fig. 6.6).

The workers at the popular education session know immediately what is happening in these paintings from the 'facial expressions', Rodriguez says. Another popular education session focusing on the company's responsibility[43] for accidents on the job revolves around a painting of an overturned bus carrying farmworkers.[44] The transportation traditionally

[41] See 'Fair Food Season 7, 2018 Update' (2018).

[42] The dumper' dumps the tomatoes from the full buckets the farmworkers bring to the truck into the large bins and is between the farmworker and the crew leader in the field hierarchy.

[43] The CIW calls the farm, the 'company'.

[44] For an image of this painting, see 'New Popular Education Drawings' (2018), https://ciw-online.org/blog/2018/08/northern-education-drawings/. One of El Teatro

Fig. 6.5 Fair Food Program Freedom from Retaliation Drawing 1. 2018. Ideas generated by farmworkers. Photograph by the Coalition of Immokalee Workers

provided for workers is over-crowded and unsafe, and the painting initiates conversations about safety and responsibility—what to do if a worker is hurt in the field and the crew leader tries to take him or her off the company property to free the company of any responsibility. After one animation, a farmworker raised his hand to share a recent experience of reckless driving that actually resulted in a case that was addressed and resolved.[45] Popular education is a cornerstone in the CIW's WSR model, and the teatros and paintings contribute to its success.

Campesino's actos, *Las Dos Caras del Patroncito*, raises this issue as well (discussed in Chapter 5).

[45] For a drawing created in preparation for 'Follow the Money March!', the Spring Action Campaign in 2020, see 'National Farm Worker Ministry' (2020).

Fig. 6.6 Fair Food Program Freedom from Retaliation Drawing 2. 2018. Ideas generated by farmworkers. Photograph by the Coalition of Immokalee Workers

Protest Art Takes on Migrant Farmscapes

The use of performance and art goes beyond popular education as it enters public discourse on human rights issues in agriculture by making the invisible workers visible not just in the art, but on the urban streets. The marches, boycotts and public fasts by the farmworkers, joined by their student and faith-based allies, present a revised farmscape where farmworkers are not commodities, but rather leaders in improving the industrial farm landscape. That revision helps to shift public opinion to a willingness, even a moral necessity, to pay more for produce in order to ensure that it is ethically sourced and to hold food giants accountable for food justice and sustainability in their supply chains, and, in turn, it causes the actual landscape of industrial farms to change. In 2000, the CIW's twelve-foot tall Lady Liberty sculpture, designed by Kat Rodriguez and built by farmworkers, joined the 230 mile 'March for Dignity, Dialogue and Fair Wage' that propelled CIW's human rights work into the public consciousness. While she is clearly a version of the Statue of Liberty, a

Fig. 6.7 Lady Liberty at the Smithsonian Museum of American History. 2019. Coalition of Immokalee Workers. Photographer: Susan Haedicke

symbol of welcome to immigrants coming to American shores, she has important differences: her raised hand holds, not a torch, but a tomato, and her tablet is replaced with a full bucket of tomatoes, her skin is brown, and the plinth on which she stands says: ‘I, too, am America’ (Klein 2018: 98–113). Lady Liberty demands to be welcomed to the United States alongside immigrants coming from Europe by ship into the New York City harbour. The farmworkers became very attached to the statue during the march. Lucas Benitez said, ‘Lady Liberty ended up being the guide, the leader, of the march’ (Klein 2018: 106). She now resides in the Smithsonian Museum of American History’s ‘Many Voices, One Nation’ permanent exhibition (Fig. 6.7).[46]

Since she was such a powerful symbol, the Lady Liberty sculpture was recreated for the public fast in front of the headquarters of Publix (‘CIW, Allies to March 150 Miles’ 2012), Florida’s largest grocery store chain, but this time carrying a bucket filled with the CIW’s graphic of

[46] See ‘Marching with Liberty’ (n.d.) for photographs of Lady Liberty, http://americanhistory.si.edu/many-voices-one-nation/new-americans-continuing-debates-1965%E2%80%932000/marching-liberty.

the glorious sun of the 'New Day' and a tablet with farmworker Fair Food Program rights. Lady Liberty and accompanying flags, signs and placards, asking questions like 'Whose hands picked your food?', 'Slavery is Everyone's Business', 'End the Poverty', 'Sweatshops in the Fields',[47] or 'Got food? Thank a farmworker', were also present at the 'March for Rights, Respect and Fair Food' in 2013.[48] Since such flags, signs and placards appear in all the long-distance marches and boycotts, it is hard for members of the public who experience this performative protest art to eat a Florida tomato without thinking about those who picked it.

For over a decade, the CIW has organised annual Spring Actions to target a particular fast food or grocery store chain that the Coalition is pushing to join the Fair Food Program. These Spring Actions, in the form of direct action marches (often over one hundred miles long), boycotts and/or public fasts lasting several days, mobilise farmworkers, students, religious leaders and the public to combat the corporation's lack of protection of human rights, and they usually cause market consequences for the corporation. For the 'Do the Right Thing Tour' in 2011 that targeted Publix, the main grocery store in Florida, David Solnit, created a colourful street pageant with large puppets. On Day 5 of the tour, three flamboyant parades started at different locations in the city and converged at Publix where they were greeted by hip-hop artist Olmeca who sang about the struggle for justice in the fields and primed the crowds for the street performance to follow. Artist Mona Caron joined Solnit to prepare the art for the march. She painted four posters with images depicting types of abuse faced by the workers in the fields. 'Slavery' shows a chained hand picking tomatoes; 'Sexual Harassment' depicts a young terrified woman trying to protect herself from the oversized hands grabbing her; 'Physical Abuse' features a large man with his clenched fist drawn back ready to punch, and 'Poverty' reveals a young man looking at his empty hands as the rows of tomatoes behind him stretch to the horizon. Each of these posters had a corresponding poster that depicted an end to each abuse: 'Respect', 'Freedom', 'Dignity' and 'Fair Wages'. The corresponding posters were placed back-to-back and transformed into four twenty-foot tall puppets with skull-like heads and open hands attached

47 This phrase was used by Edward R. Murrow in *Harvest of Shame*.

48 See a video of the new liberty statue being carried in the march, https://www.youtube.com/watch?v=1zFWy11EPUM (also on 'A Look Back' 29 March 2013). See also 'Social Media Chatter' (2013).

to the tops of the posters.[49] They became the core of the event in front of Publix in Tampa, Florida: a street performance in the form of a *canta-historia* (story told or sung and accompanied by visuals).[50] The street performance began as dozens of farmworkers pushed their way through the crowd and turned the area in front of the stage into a tomato field as they mimed picking tomatoes and lifting the full tomato bucket onto their shoulder to carry to the truck to be dumped. Behind them on the stage, four farmworkers gave testimonies of abuses they had experienced. As they narrated the history of the CIW, the large puppets depicting the abuses towered over them. In the performance, as the farmworkers and their allies[51] began to win victories, in spite of Publix's refusal to join the Fair Food Program, one-by-one the puppet depicting an abuse turned to reveal the corresponding solution (e.g. from slavery to freedom) to transform the world into one of justice and dignity. The participatory performance invited spectators to contribute to the narrative and join in celebrating the moment of certain victory. Like so much of the CIW's protest art, this performance played with collage tactics, described by Jacque Rancière (discussed in Chapter 5) not only with the juxtaposition of the worlds of injustice and justice, but also, in a more complex form, with the testimonials exposing real abuse speaking in juxtaposition to a large placard-puppet on the actual street in front of Publix, the corporation responsible for the abuse in their food chain. Fiction here does not work in opposition to reality; rather the imaginary re-frames, re-interprets and overturns the real to locate the inexorable future tangibly in the less certain present.

In 2013, David Solnit and Mona Caron again joined forces for the 'March for Rights, Respect and Fair Food' Spring Action to create another street theatre piece, this time a *mística*, a non-verbal allegorical play with its origins in liberation theology and Latin American popular theatre. Brazil's Landless Workers' Movement (Movimento dos

[49] Photographs of the poster puppets in the street performance and a description of the play are available at Caron (2011a), https://monacaron.com/artivism/coalition-immokalee-workers-visuals-mass-action-floridas-farmworkers, Caron (2011b) and 'Do the Right Thing: Day 5 Update' (2011). See also 'With the Coalition of Immokalee Workers—Visual for Mass Action' (2011).

[50] See Solnit performing a canta-historia on different topic that appears at the end of his videotaped presentation at the conference, *Artists Now!* (Solnit 2016).

[51] See the Alliance for Fair Food website.

Trabalhadores Rurais Sem Terra), well known to the CIW, uses the *mística* extensively in its cultural activism.[52] The CIW *mística* follows the Brazilian model of depicting the abuses and affirming they can be overcome as it tells its story of the transformation of industrial farmscapes through the Fair Food Program since 2000 and predicts success to get Publix to sign on to the program. The Coalition performed the *mística* in front of Publix's Headquarters in Lakeland, Florida.[53] Its primary image was a large cardboard cut-out of a flat-bed truck carrying the 'workers' rights' depicted as large cardboard tomatoes, each representing one of the farmworkers' demands for better working conditions, like One Penny More, Minimum Wage, Work Free from Sexual Harassment, or To Report Abuse without Fear. The truck was stuck in a ditch of years of farmworker exploitation before the Fair Food Program. The workers mimed the history of the CIW's Campaign for Fair Food: a solo farmworker tried unsuccessfully to push the truck out of the ditch, so he asked for help from other workers and then made alliances with student and faith-based groups to push the truck with its important cargo out of the ditch. Still no success, so they began their negotiations with the fast-food chains and grocery stores. In the street performance, as each one signed on, the 'corporation' (represented by a recognizably costumed performer, i.e. the Taco Bell chihuahua or Ronald McDonald) joined the farmworkers to push the truck out of the ditch. Only Publix refused to join, trying to keep the truck in the ditch with his foot pushed against the front fender. Yet the CIW farmworkers and their allies, joined by members of the public, gave a final push and got the truck back onto the road. Lucas Benitez commented, 'It's not a question of whether we will win, but when. And when we do win, we will not only help free workers from oppressive conditions in the fields, but we will also free Publix from the impossible burden of supporting and justifying that oppression'.

In addition to street theatre, the CIW uses ordinary objects elevated to *metonym* for the farmworkers' counter-narrative of dignity, respect and

[52] See Flynn (2013) and Hammond (2014).

[53] For photographs, see Caron (2013b), https://www.facebook.com/mona.caron/media_set?set=a.10200717914091569.1073741831.1273193655&type=3. See also Caron (2013a), 'March for Rights, Respect and Fair Food: Day 15 Photo Report' (2013), http://www.ciw-online.org/march/updates/march_17.html?fbclid=IwAR3Ht3dDDV6vcAIeD_UkwN5vZdDBmOXxb3b2T8Kt9vkYrAgBNlcZf3pWBnQ, 'With the Coalition of Immokalee Workers, Sets for a mística' (2013), and 'Theatre, Reflection Make for a Festive Evening' (28 February 2013).

justice. Here the object goes beyond a stimulus enabling comprehension of a situation as it also elicits an affective response as the object stands in for the farmworker's body. The CIW's protest art plays with that idea in various ways. One object is the iconic tomato bucket. During the Taco Bell campaign in 2004, farmworkers repeatedly created a pyramid of tomato buckets over two stories tall to represent a day's work for one farmworker: a shape that evokes a pinnacle and strength.[54] These are not necessarily new, clean buckets, but rather ones still bearing sticky traces of their time in the fields, traces of tomatoes and pesticides and soil, and so exude a reality through smell and touch that complicates their identity as a symbol—or rather enlarges it. Also, a 'full' tomato bucket weighing thirty-two pounds often acts as a prop in information sessions for the general public. As a recognizable object, especially in the Florida context, the 'working' tomato bucket escapes its 'thing-ness' and provokes an experiential engagement. As Nely Rodriguez explains, the bucket represents the reality of the Florida tomato pickers and is an iconic symbol of the Florida farmscape (Personal interview, 5 October 2018).

Other objects join the CIW repertory of symbols. During the Burger King campaign in 2008, one of the Burger King executives aggressively challenged the poverty of farmworkers saying that they could actually make $18/hour if they just would 'pick fast'—a comment that offensively drew on a Mexican stereotype prevalent in the United States by blaming the 'lazy' workers for their poverty and thus absolving the giant retailers of any blame. To earn eighteen dollars an hour would actually require a worker to pick an impossible forty buckets (1280 lb/580 kilos) every hour (the average for a worker picking at full speed is fifteen to twenty buckets an hour). In the march to the Burger King headquarters in Miami, the farmworkers contradicted the false accusations by creating a 'Walk in Our Shoes' action. The CIW asked farmworkers to donate their used shoes from work, shoes not cleaned, but as they were when they left the fields—sweaty and dirty from the sandy soil and green tomato residue. The donated shoes were old, torn, filthy and smelly. The marchers pulled a wagon full of their old shoes with a banner on the side that challenged: 'If you doubt our poverty, walk in our shoes' and so offered a counter-narrative as a challenge to the Burger King management to think about the work that the farmworkers do every day, to take a moment to 'walk

[54] For a photograph, see 'CIW Protest' (2004), https://ciw-online.org/blog/2004/05/ciw-protest-at-yum-brands-shareholder-meeting/.

in our shoes'. In other campaigns, they have created actions that 'Expose the Dirty Laundry' of the large food retailers as they hang their sweaty and stained t-shirts in front of the corporate headquarters to emphasize the retailer's responsibility for their supply chains and thus the farmworkers' poverty. The sight and smell of the old clothing worn by farmworkers working in the fields, many of whom are there in person on the march, propel the public and sometimes the corporate management to abandon a distanced position of objective observation. The object is no longer just a thing to look at, but rather it fosters a relationship with the spectators. As Nely Rodriguez explained to me: 'That is a really powerful tool—just by looking at some shoes, seeing and smelling some t-shirts, you can take a moment to think about the kind of hard work that person had to complete. And the physical energy that is wasted throughout the day, the physical energy that you have to come up with can be felt and seen just by looking at a pair of shoes, at a t-shirt' (Personal interview, 5 October 2018). 'Objects' also have an online presence in a video called *Con Estas Manos/With these Hands* (2007) that offers a collection of photographs of the pickers' hands as they finish a day's harvesting of tomatoes—hands stained green and brown from the tomatoes, hands holding tomatoes, hardworking and gnarled hands. The opening words declare 'with these hands I demand the future that poverty wages have stolen from me'. As Greg Asbed told me, this focus on ordinary objects so essential in tomato-picking enables the public to understand the hard labour involved in this kind of work through what the farmworker uses each day—T-shirts, shoes, tomato buckets and hands. These object icons, ubiquitous in the CIW's protest art, acquire the status of 'symbol' for complex ideological positions in the US agricultural sector. The object-symbols in the CIW's temporary public art 'materialise' very specific aspects of the tomato pickers' working conditions. They remain intimately linked to their referents, so their power to create an alternative industrial farmscape in the public imaginary—an imaginary that can then force a transformation in the actual landscape—comes from the emotional and somatic responses that the objects arouse in their viewers rather than from their symbolic representation of or substitution for the issue they symbolize. As the viewer beholds an object like this with a reality beyond the ordinary object, it transforms a food *consumer* into a food *citizen* and begins to contribute to the process of structuring a new agribusiness landscape.

The strategy of raising the pedagogical and somatic power of an object from *thing* to *experience* escalated in the CIW's two mobile museums—the Florida Modern Slavery Museum (2010) focusing on forced labour of farmworkers and Harvest Without Violence Mobile Exhibit (2017) on sexual harassment and rape in the fields. Both mobile museums were created by farmworkers. As temporary art exhibition sites in public spaces, these museums and their carefully curated objects acquired an aura of almost reverent special-ness. The mobile museums' exhibition spaces themselves evoked the original sites where the abuses took place, and that locational specificity enhanced a sense of participatory witnessing for the museum-goer, propelling him or her to *see*, viscerally *experience* and *acquire* a critical consciousness of the events. These exhibits, like the shoes and t-shirts, offered more than archival evidence of atrocities. In her work on the 'memory paths' of the performances of the Peruvian theatre company Yuyachkani, Diana Taylor explains that they 'enter into a dialogue with history of trauma without themselves being traumatic. They are carefully crafted works that create a critical distance for "claiming" experience and enabling, as opposed to "collapsing," witnessing' (Taylor 2003: 210). This insight rings true in the performative mobile museums as well. As with the performances Taylor describes, the Florida Modern Slavery Museum and Harvest without Violence Mobile Exhibit encouraged viewers to 'recognize their role in the ongoing history of oppression which, directly or indirectly, implicates them. ...They teach the communities not to look away' (Taylor 2003: 211).

The Florida Modern Slavery Museum was part of the CIW's Anti-Slavery Program and documented both the enforced labour suffered by these workers and the Coalition's investigations that uncovered and helped prosecute twelve cases of slavery on farms in the southeast of the United States, freeing over 1200 slaves. In 2009, a shocking news story of the Navarrete family who held twelve tomato pickers as slaves on their property in Immokalee surfaced. At night, the migrant workers were chained in windowless cargo trucks with no ventilation or toilet facilities and only released to work in the fields on neighbouring farms during the day. They were threatened with beatings and death if they tried to escape. One night, three of the men broke through the damaged roof of one cargo truck, freed the other men, and immediately reported their enforced confinement to the CIW and the Collier County Sheriff. Although there had been other slavery cases investigated by the CIW (six between 1997 and 2007), this case only forty miles from the city of Naples catapulted the issue into the public's consciousness. In addition, newspaper coverage in the *Fort-Meyers News Press* explicitly linked local growers and Publix to slavery as the Navarretes took their prisoners to harvest tomatoes on the local farms that sold to Publix. The Florida Modern Slavery Museum documenting this case and others was set up in a cargo truck, an exact replica of the one involved the Navarrete case, so visitors stepped into an alternate reality as they crossed the threshold to enter the airless, darkened space—not a comfortable museum-going experience, but one that suggested what the Navarrete slaves experienced. Inside were chains that bound them and many other objects found on site as well as testimonials and information about the other slavery cases. Outside was information about modern-day slavery, particularly of migrant workers. The CIW's Anti-Slavery Program has been so successful that other cases of enforced labour or human trafficking in agriculture have not been identified in the Southeast United States until recently ('Latest Slavery Indictments' 2021). This mobile museum, that visited many cities in the eastern half of the United States, has not been exhibited recently and has been partially dismantled.

The Harvest without Violence Mobile Exhibit[55] presented an equally frightening agricultural landscape on many industrialised farms—a farmscape rife with sexual harassment and rape. On a national level, women experiencing sexual assault at work is unacceptably high at approximately 25%, but in the fields, 80% of women experience sexual assault (*Food Chains*).[56] Zero tolerance of sexual harassment has been one of the core principles of the Fair Food Program from its beginning in 2011, but when the huge fast-food chain Wendy's dropped their long-standing Florida tomato suppliers who had joined the Fair Food Program and began buying tomatoes from Mexico where sexual harassment of women farmworkers is well documented, the farmworkers decided to respond with a campaign and a mobile museum. In 2017, the CIW's Women's Group, El Grupo de Mujeres, launched the Harvest Without Violence Campaign as part of the larger campaign to convince Wendy's to join the Fair Food Program. They sought to foreground Wendy's complicity in violence against women and to offer customers a choice of participating

[55] See also 'Can't Miss Radio' (2017) and 'After #MeToo' (2020).

[56] See the second video on Alejandra Cerrera on CNN Freedom Project (2017a) and Rape in the Fields (2013).

in that violence or boycotting Wendy's in protest until it became part of the Fair Food Program. The Group argued: 'If we don't speak up, we give up, and we will never give up. We are building a path of respect and dignity for ourselves, for our daughters, for all workers' ('Farmworker Women' 2017). The Harvest Without Violence Mobile Exhibit was part of this campaign, and it educated the public on the hidden reality on farms outside the Fair Food Program and Wendy's choice to support violence and impunity over human rights.[57] The exhibition space is an octagonal tent thirteen feet in diameter that creates a feeling of entrapment, a sense that the visitor is enclosed, surrounded—a feeling that many women experience when working in the fields. Nely Rodriguez says it could be one woman alone in this rural space with no way to get out or a group of women 'herded' into a space of fear and violence. For Rodriguez, it is this 'climate' that the museum-goer steps into, a climate that farmworker women battle every day. Inside the tent is what Marley Monacello, CIW staff member who curated the exhibition with women farmworkers, calls 'Chapter 1' of the story. Here, the hidden landscape of abuse on farms outside the Fair Food Program is revealed. Oversized images of rows of tomato plants as far as the eye can see, of women getting onto the buses to go to the fields and harvesting crops in the fields cover the walls establishing a sense of isolation, fear and danger. Testimonies of women's experiences in the fields of Mexico and the United States, information on the extent of sexual harassment on farms in the two countries, multi-media presentations, articles by investigative journalists and scholars, and court documents reveal the magnitude of the problem. The brutal experiences endured by the women are palpable. Escaping the claustrophobic atmosphere of the tent, visitors step outside to experience the very different farmscape of safety and justice for women on farms participating in the Fair Food Program. The first set of panels that follow the outside of the tent make up 'Chapter 2'. Here, visitors read about real-life successes accomplished by the Coalition's Fair Food Program and its Worker-driven Social Responsibility, including popular education sessions, monitoring by the Fair Food Standards Council and meaningful market consequences for the retailers and growers if they

[57] For photographs, see 'Starting Today' (2017), https://ciw-online.org/blog/2017/11/harvest-without-violence-nyc/ and 'Part One: CIW's Women's Group' (2017), https://ciw-online.org/blog/2017/10/part-one-ciw-womens-group-harvest-without-violence-tour-arrives-in-columbus/.

ignore violence in the fields. The rest of the panels, 'Chapter 3', offer information on the on-going Wendy's campaign and Wendy's position on the issue. The fallacies of Wendy's arguments are clearly on display. At the end of the panels is a quilt, six feet by five feet, created from fabric squares, all the same size. On each square, a woman farmworker has handwritten her thoughts and emotional responses to physical assaults in the fields as 'messages to Wendy's' about their experiences and fears. These messages, in Spanish, English and Haitian Creole, create emotionally raw personal testimonials. Stepping back from the quilt, one can see that the coloured squares suggest a scenic representation of the agricultural landscape itself with greens and browns at the bottom, greys in the middle and blues with birds toward the top. But the scenic representation also suggests the trajectory of the Fair Food Program's impact on tomato field 'landscape' as the danger on the ground shifts to blue skies once the program is implanted (Fig. 6.8). The exhibit performs the transformation of the industrial farm landscape in the southeast United States from one

Fig. 6.8 Quilt-making for Harvest Without Violence Mobile Exhibit. 2017. Photograph by the Coalition of Immokalee Workers

of violence and exploitation to one of justice and dignity, and it reminds viewers that there is still work to be done.[58]

Hunger and Harvest: CIW's Public Fasts

The actual bodies of the farmworkers have a constant presence in the industrial farm landscape in the fields certainly and also in the various forms of protest art that perform these farmscapes, but their most powerful performative presence is in the public fasts that act as performances of resistance. Following in the footsteps of the UFW, the CIW has engaged in numerous hunger strikes and 'rolling fasts'[59] as a way to tell their counter-narrative of human rights, respect and fair food and to place themselves in that landscape. The public fast held in 2012 as part of the campaign to convince Publix, the major supermarket chain in Florida, to join the Fair Food Program, is the focus of the documentary *Food Chains* that explains how the market power of large retailers to push down food prices results in sub-poverty wages for the farmworkers. By 2008, the CIW had drawn attention to Publix's complicity in farmworker abuse when it presented proof that some tomatoes sold at Publix had been sourced from a grower prosecuted in the Navarrete slavery case. The CIW co-founder, Lucas Benitez, was scathing when he wrote, 'Publix likes to cast itself as Florida's community grocer—the good neighbour. But how can you be a good neighbour when people are… forced to work as slaves … in your own backyard and you turn a blind eye?'[60] The revelations, however, were not enough.[61]

The six-day 'choreographed' public fast, 5–10 March 2012, in front of Publix's headquarters in Lakeland, Florida became a performative event that focused all the damaging information on Publix into a story written

[58] Marley Monacello described this Harvest Without Violence Mobile Exhibit in an informal conversation with the author in Immokalee on 14 February 2019.

[59] In rolling fasts, a participant ends a fast after a limited number of days and it is picked up by a new faster, so the entire fast can last for months. Cesar Chavez's last fast did this as discussed in the previous chapter. In 2017, students supporting 'Boot the Braids' campaign to convince universities to ban Wendy's from their campuses, participated in a rolling fast that spread to many university campuses simultaneously.

[60] Quoted in Marquis (2017, 155–156).

[61] Perhaps *The Wall Street Journal* revelations of the link between Publix and the 6 January 2021 insurrection (mentioned earlier) will make a difference ('Publix: Radicals' 2021).

with and on farmworkers' hungry bodies, bodies made visible and vocal, bodies demanding to be treated with respect and dignity for their essential work, bodies rejecting the label of 'commodity', bodies performatively revising the industrial farm landscape. As a farmworker said on the first day of the fast, 'we workers don't have the money to pay millions of dollars to advertise in the same way that Publix advertises on TV or the radio. But we do have our bodies that we can use to send out our message' (*Food Chains* 2014). The hunger strike with its focus on the hungry body not only draws attention to a lack of human rights as represented by a lack of food, but it creates a performance of resistance against those responsible for the injustice. The fast was performed by the farmworkers for an audience of Publix customers and executives, but that actor/spectator binary soon broke down. The 'performing' of the farmworkers' story through the public fast enabled customers not just to experience it vicariously but to participate in it, creating an 'experience of immersion or "transportation"' (Polletta 2008: 27) that, in turn, contributed to a shift from ignorance of the farmworkers' plight to determination to buy food ethically harvested: a journey of food citizenship. Even some of the people working at Publix Headquarters who had to drive past the fasters on their way to and from work each day developed a sympathetic bond as many shifted from looking straight ahead, 'not seeing' the farmworkers on the first day to honking, waving and shouting words of encouragement as they passed five days later. Some even participated in the performative activities or vigils.

The public fast took place in the public space in front of Publix headquarters and thus located the farmworkers, not in the hidden landscapes of industrial farms where abuse and exploitation could be invisible, but in the open where their grievances and poverty became visible. In addition, the fast shifted the meaning of the site from corporate power to farmworker resistance to that power. It began with an opening ceremony of speeches and lighting the flame that would remain lit throughout the fast. The thirty initial fasters, all farmworkers, picked up their placards saying things like, 'I go hungry today so my children won't have to tomorrow', 'Whose hands picked your food?, 'I too am human!' and 'Publix, be a good neighbour' and took their places along the entrance drive to Publix's headquarters. They stood still and solemn as employees arrived for work. The day of fasting was punctuated with speeches, education sessions, poetry reading, prayer circles and music, and ended with a candlelight vigil at one of the flagship Publix stores in Lakeland. The words of Reverend Rankin, one of the speakers at the vigil, vividly depicted links between families buying tomatoes *at* Publix and families picking tomatoes *for* Publix. The next day, Publix brought in police to make sure things remained peaceful, but the fasters did everything according to the law, so the police just watched the spectacle of the fast. Publix employees too watched the farmworkers fast as they ate their lunches on the lawn across the pond from the fasters, and the fasters, in turn, watched them eat. This 'mirror' spectatorship raises questions about the narratives each group saw and what they understood. Nely Rodriguez discussed the sense of justice, hope and solidarity that developed among the fasters through various activities like their feet-washing ceremony or the 'calling out' of abuses that had occurred in the fields as well as from support for the farm workers with honks, waves and more from the Publix employees. Is that the narrative that the employees saw? For some, probably yes as there were visible changes in their reactions to the fasters, but others probably saw an opposing narrative. Each day for the next five days, from sunrise to sunset, looked similar except the number of fasters grew to hundreds as members of the public who supported the farmworkers' cause joined the fast (Rodriguez, Personal interview, 5 October 2018). As food citizens, they walked with farmworkers carrying the recognisable tomato buckets around the headquarters and wrote hundreds of notes to Publix and tied them to the chain link fence surrounding the site. Forty-four years to the day that Cesar Chavez broke bread with Robert Kennedy to end his fast for farmworker justice in 1968, the CIW ended their fast with a ceremonial breaking of bread with Kennedy's widow, Ethel and his daughter Kerry.[62]

As a performance of dissent and agency, the public fast enabled the farmworkers most severely affected by the policies of supermarket chains like Publix to expose the landscape where they pick tomatoes and to bring their issues into the public consciousness. But more significantly, it enabled the farmworkers to cast off the identity of helpless victims or commodities and to reframe themselves as self-respecting and committed human beings who put food on American tables, even though they do not earn enough to put food on their own family's table, and it also

[62] For a short video of the bread-breaking ceremony of the fast, see 'Support for the Fast, Campaign, Wide, Deep' (2012), https://ciw-online.org/blog/2012/03/fast_support_post/.

engaged the public by propelling them into assessing their ethical values. Through the public fast, the farmworkers not only constructed a new and compelling identity of a farmworker 'David' battling the 'Goliath' of Publix for justice, but also watched their David grow as members of the public joined the fasters. Their fasting 'disappearing' bodies became the metonym for their invisible labour in the food chain, but, in contrast as in a collage, their daily presence outside Publix headquarters attested to the moral obligation to make visible their essential role in the agri-food system. The visibility of the farmworkers' physical need for food *not* being met during the fast was linked to their less visible lack of toilets, clean water to drink, and shade in the fields that can often surpass 90 °F (about 33 °C). These unfulfilled necessities in the fields became the core of the performance event away from the fields. The fasting site became a political space in which the spectacle of their hunger made visible their demands and the conditions around those demands as the *lack* of food acquired a material presence and contributed to a revised migrant landscape through its endorsement of food citizenship.

Performing the Migrant Farmscape Online

The protest art on the street brings the agribusiness landscape into the physical public sphere as it seeks to tell an effective and affective story that responds to the changed economic-agricultural landscape, but the CIW's counter-narratives in online videos and in its social media campaigns have an impact further than the street and have proved most effective in tarnishing the brands of the large retailers. A short information video entitled One Penny More (2010) created as part of the social media campaign to launch the Fair Food Program, depicts the power of collage to blur the sharp divisions between land and labour and between farmworker and consumer by linking the tomatoes in the fields filled with overworked, underpaid and under-protected tomato pickers to the tomatoes on the supermarket shelves. That collage 'reveals one world hidden beneath another: capitalist violence behind the happiness of consumption' (Rancière 2006: 87). The video acts as a stimulus for dialogue and change from consumer tolerance to citizen activism. The video's cinematography relies on rapidly shifting points of view and juxtaposition to write the customer, farmworker and retailer into one narrative. Elmer Bernstein's familiar 'vacation' tune, 'A Summer Place' begins as we join a shopper filling her cart in a spotless grocery store. She focuses on her shopping, not on the story of the fruit's journey to the grocery store, but as she enters the lush produce aisle, the image morphs from green lettuces displayed on the shelves to hot and dirty tomato fields as workers race to fill their buckets. The shift in point of view from supermarket to field begins to establish an enduring connection between shopper and farmworker that is repeated several times: for example, the farmworker rushing to pick as many tomatoes as possible as fast as possible is contrasted to the shopper carefully choosing the best-looking tomatoes or the shopper lifting a small melon fades into the farmworker lifting the thirty-two-pound bucket full of tomatoes. The video also draws attention to the accountability of the supermarket in worker exploitation not only in the fields, but also in the stores as a farmworker carrying a bucket of tomatoes on his shoulder dissolves into a stocker carrying a crate of tomatoes on his shoulder. This fade underscores the parallel between the supermarket's low wages for stockers and the industry's low wages for farmworkers, both at the bottom of the food chain. Another scene contrasts what a worker is paid for one pound of tomatoes and what a shopper pays for the same amount. As the camera zooms in on the price of $2.99 per pound of tomatoes, the words 'For every pound picked, workers are paid 1.4 cents' appear. The video ends with the shopper willingly paying a penny more for her tomatoes to support the Fair Food Program's demand for one penny more for every pound picked to go to the pickers. Here we see the contrasting narratives: the agribusiness story of an abundance of perfect tomatoes grown under the warm Florida sun that magically arrive on the supermarket shelves at a low cost as opposed to the hidden world of the exploited workers sweating to harvest the crops and struggling to feed their own families: one world beneath another. The video effectively raises public awareness and so acts an essential tool to achieve food citizenship, but it is clear that the general public is not the only audience for the CIW's art. To achieve fair pay and better working conditions, audience awareness must be paired with the retailer's willingness to act, so the other audience for this protest art is the management of the corporate retailers as the CIW uses the art to undermine their brands. The efficacy of strategies such as these is evident in the fact that so many multi-billion dollar retailers recognize the power of the CIW's counter-narrative and sign on to the Fair Food Program.

In The Story Behind Your Food (2018), released as part of the Wendy's campaign, the inseparability of the world of the shopper and the world of the farmworker and the justice afforded by Fair Food Program is again the focus. *The Story Behind Your Food* has earlier iterations. An online post entitled 'A Tale of Two Roadside Farm Stands'[63] presents a photograph of a bountiful farm stand in front of verdant fields and asks the reader to engage in a brief 'thought experiment'. It asks what if you needed fresh tomatoes, and you drove past this idyllic farm stand full of fruits and vegetables freshly-picked in the surrounding fields? It assumes that you would stop, but as you approach the cash register with your tomatoes, the world hidden from shoppers comes into focus when you see a farm boss beating a farmworker who pleads with him to stop saying he only wanted a drink of water. The thought experiment continues: 'Stunned and concerned for the worker's wellbeing, you ask the attendant to intervene, but he just looks at you, smiles again, and says, "Oh, that? That happens all the time. Nothing to worry about. That'll be $9.50". What would you do?'. It then offers an ethical choice: Be like Wendy's and ignore the abuse for low-cost tomatoes? Intervene? Or buy tomatoes from the farm stand across the road that cost pennies more but are sourced ethically. This thought experiment exemplifies the way art can construct a path to food citizenship as it encourages the reader not just to focus on the abuse but to join a 'dialogic collaboration' and become part of the solution by boycotting retailers with unethical practices in their supply chains, like Wendy's. At the end of that post is a drawing that shows two fields connected by a bridge. On the one side of the bridge is an industrial farm landscape before the Fair Food Program with the field under a stormy sky and where a boss beating a farmworker while the recognizable little girl who represents Wendy's smiles and turns her back. On the other side of the bridge is the industrial farm landscape transformed by the Fair Food Program where the sun is shining as the workers harvest tomatoes without fear of violence ('Join the Return to Human Rights Tour!' 2017).

The CIW also has run social media campaigns to further its human rights activities. Inspired by Dr Seuss' *How the Grinch Stole Christmas*, Student/Farmworkers Alliance (SFA) led a campaign called 'Wendy's the Fast Food Grinch' in 2016. The CIW and SFA asked consumers to post 'Wendy's the Grinch' videos online to 'change the heart of a penny-pinching, mean-spirited scoundrel' ('Post a Video' 2016). The videos with remixes, performance skits, staged conversations, monologues and

[63] 'A Tale of Two Roadside Farm Stands' (2017). https://ciw-online.org/blog/2017/02/a-tale-of-two-roadside-farm-stands/.

staged 'break-up scenes' where a person confirms his or her love for Wendy's but must end the relationship because of Wendy's unacceptable behaviour flooded the SFA FaceBook page.[64] This campaign significantly increased the numbers of students, especially on university campuses where Wendy's has a strong presence, participating in the boycott as hundreds of videos were uploaded. More recently, the CIW cancelled the Spring Action 2020 'Follow the Money' march to Wendy's Headquarters in New York City, scheduled 10–12 March 2020, because of the coronavirus pandemic. But with the cancellation of the march on the streets of New York City, the CIW launched a social media campaign targeting the top shareholders who have influence over Wendy's decision-makers: 'since we can't be in the streets this week, we will need every single person in the Fair Food Nation … to help carry our voices through the *virtual* streets of social media, and into the suites and boardrooms of the deep-pocketed investors behind Wendy's' ('Take to Social Media to Urge Wendy's Top Investors' 2020).[65] They sought to inundate Wendy's investors and banks, including Goldman Sachs, BlackRock, Wells Fargo and BNY Mellon, with posts demanding that they convince Wendy's to join the Fair Food Program. Mimicking the way that the CIW shifted their focus from the growers to the grocery store and fast-food chains in the early twenty-first century, this call goes further up the food chain to expose the complicity of the investors. It will take time to know the success of this social media campaign, but at Wendy's most recent annual meeting in May 2021, the shareholders voted overwhelmingly (95.28%) for transparency in Wendy's supply chain. Signing on to the Fair Food seems inevitable.[66]

The CIW's art-based storytelling of counter-narratives seeks to politicize the consumers by exposing abusive labour practices hidden from the consumers' view and the retailers' complicity in that abuse and to shame retailers into taking responsibility for an ethical food chain. The CIW's narrative in its multiple forms revolves around concepts of dignity, dialogue, justice and empowerment through food citizenship. While the setting of their stories exposes the exploitation, the characters challenge a victim narrative. Such protest art interrupts the normative agricultural

[64] See 'Video Action: Boycott the Fast Food Grinch!' (2016).

[65] See also 'Take to Social Media: #BOYCOTTWENDYS' (2020).

[66] See 'Massive Display of Support' (22 May 2021).

narrative by staging an alternative present and shaping a different future. Clearly, without Worker-driven Social Responsibility in the Fair Food Program, its Code of Conduct and the Fair Food Standards Council that monitors adherence to the Code of Conduct on participating farms, the art would not have achieved the many successes or revised the agribusiness landscape. Yet without the art, CIW would not have acquired what David Solnit calls a 'people power strategy' (Renz 2005: 2) that helps change opinions and behaviour. The CIW has effectively wedded performative art and direct action skills to imagine a changed landscape and to give people hope that it is achievable and tools to achieve it. And even during the pandemic, the use of performative protest art to clarify the political message is powerful, as in the May Day Virtual Conference for Farmworkers in 2021 that featured musical performances of activist-artists Tom Morello, Olmeca, Mare Advertencia Lirika, Son Jaorcho collective Lxs Altepee and comedian and actor Amy Schumer interspersed with political messages from fellow organisers in the United States and Mexico.[67] The art and the activism work together to perform the revised landscape in the protest art and in the fields. Identified cases of modern-day slavery are now rare in the agricultural sector in much of the United States, cases of sexual harassment in the fields on Fair Food farms have dropped significantly, the CIW's WSR model to achieve human rights in agriculture is being replicated in other industries, and the CIW farmworker leaders train activists in WSR beyond the US borders. Nevertheless, the CIW's work is far from finished. While the public fasts brought visibility to the farmworkers, they have not yet succeeded in getting Publix's executives to the negotiating table. As this book goes to press, that campaign, as well as one against the fast-food chain Wendy's, are ongoing. At the same time in the summer of 2020, Smoky Mountain Family Farms joined the Fair Food Program to add Tennessee to the list of states where the FFP is active ('Fair Food Program Expansion Marches On' 2020), and Bloomia, the largest cut flower farm on the East Coast also became a participant ('Largest Cut Flower Farm' 2020). In early 2021, CIW created the Fair Food Sponsor Programme to enable co-ops, independent grocers, restaurants and chefs to join the Fair Food Programme, and in June, Each Peach Market in Washington, DC was the first independent grocery store to sign up. ('Announcing' 3 June 2021). The pairing of protest art and

[67] See 'In Case You Missed It, Wendy's Boycott Goes Global' (7 May 2021) for videos of the virtual concert.

direct action of the Coalition of Immokalee Workers shows how efficacy is viable in art-agriculture partnerships as the CIW achieves verifiable improvements in the industrial farm landscape and has shifted many food *consumers* into food *citizens*. The Fair Food Program offers a model for food democracy.

Works Cited

'2019 in Review, Part 4: The Vast Potential of Worker-driven Social Responsibility'. December 31, 2019. [Blog] CIW. https://ciw-online.org/blog/2019/12/2019-in-review-part-4/.

'4 for Fair Food Tour' 2019. [Blog] CIW. https://ciw-online.org/?s=4+for+Fair+food.

'Action Round-Up: Farmworkers, Students and Religious Allies Rally for Farmworker Justice in Sarasota, Philadelphia and Miami!'. 30 November 2018. [Blog] CIW. https://ciw-online.org/blog/2018/11/action-round-up-nov/.

'After #MeToo, This Group Has Nearly Erased Sexual Harassment in Farm Fields'. 13 March 2020. [Blog] CIW. https://ciw-online.org/blog/2020/03/vera-chang-civil-eats/.

Alliance for Fair Food. http://www.allianceforfairfood.org.

'The Alliance for Fair Food: Uniting for a New Day in the Fields'. 2018. [YouTube video] https://www.youtube.com/watch?v=SYfKFmGNghQ&t=178s.

Angel, Jen. 2008. 'David Solnit and the Arts of Change'. www.joaap.org/webonly/solnit_angel.htm.

'Anita Hill's Hollywood Commission Announces Collaboration with the CIW, Fair Food Program!'. 25 May 2021. [Blog] CIW. https://ciw-online.org/blog/2021/05/hollywood-commission-partnership/.

'Announcing the First Fair Food Sponsor in the Nation's Capitol!'. 3 June 2021. [Blog] CIW. https://ciw-online.org/blog/2021/06/announcing-the-first-fair-food-sponsor-in-the-nations-capital/.

'"Because a Tractor Does Not Tell the Farmer How to Run the Farm": Testimony by Lucas Benitez'. 15 April 2008. [Blog] CIW. https://ciw-online.org/blog/2008/04/lucas_senate_testimony/.

Bacon, David. 2002. 'No Quiero Taco Bell! David Bacon Interviews Lucas Benitez of the Coalition of Immokalee Workers'. *In These Times*. https://inthesetimes.com/article/no-quiero-taco-bell.

Bruzzi, Stella. 2006. *New Documentary*. 2nd ed. London and New York: Routledge.

'Can't Miss Radio: Public Radio International Story Brilliantly Captures Fair Food Program Success in Fight Against Sexual Harassment in the Fields!'.

17 November 2017. [Blog] CIW. https://ciw-online.org/blog/2017/11/pri-diocese/.

Caron, Mona. 2011a. 'With the Coalition of Immokalee Workers—Visuals for Mass Action by Florida's Farmworkers'. https://monacaron.com/artivism/coalition-immokalee-workers-visuals-mass-action-floridas-farmworkers or https://monacaron.com/artivism/coalition-immokalee-workers-visuals-mass-action-floridas-farmworkers.

Caron, Mona. 2011b. 'Do the Right Thing'. https://www.facebook.com/media/set/?set=a.1859178880548&type=3.

Caron, Mona. 2013a. 'With the Coalition of Immokalee Workers—Sets for a Mística Theatre at the March for Rights, Respect and Fair Food'. https://monacaron.com/CIW-mistica.

Caron, Mona. 2013b. 'CIW 200 Mile March Across Florida for Rights, Respect and Fair Food'. https://www.facebook.com/mona.caron/media_set?set=a.10200717914091569.1073741831.1273193655&type=3.

'CBS Sunday Morning: Fair Food Program'. 2015. [TV programme] https://www.youtube.com/watch?v=b6s4YAOISns.

'CIW, Allies to March 150 Miles to Lakeland in March for Rights, Respect and Fair Food!'. 2012. [Blog] CIW. http://ciw-online.org/blog.2012/12/spring-action-announced/.

'CIW's Greg Asbed Named 2017 MacArthur Fellow'. 19 October 2017. [Blog] CIW. http://ciw-online.org/blog/2017/10/greg-asbed-macarthur/.

'CIW Protest at Yum Brands Shareholder Meeting'. 20 May 2004. [Blog] CIW. https://ciw-online.org/blog/2004/05/ciw-protest-at-yum-brands-shareholder-meeting/.

'Climate Underground'. 13 November 2020. Hosted by former Vice President Al Gore. [YouTube] https://www.youtube.com/watch?v=dviARmmI1NE.

CNN 'Freedom Project'. 2017a. https://edition.cnn.com/2017/05/30/world/ciw-fair-food-program-freedom-project/index.html .

CNN 'Freedom Project—Wendy's Boycott'. 2017b. [TV programme]. https://www.youtube.com/watch?v=OKm_yZvBOJ0.

Coalition of Immokalee Workers. https://ciw-online.org/.

'Coalition of Immokalee Workers: La Mixteca—Street Theatre. 2003 Taco Bell Truth Tour'. 2003. [YouTube]. https://www.youtube.com/watch?v=bB1zNasI_Dc.

Con Estas Manos/With these Hands. [YouTube] CIW video. https://www.youtube.com/watch?v=ItRPBp8kj8o.

Dixon, Beth A. 2018. *Food Justice and Narrative Ethics*. London: Bloomsbury.

'Do the Right Thing: Day 5 Update'. 2011. http://www.ciw-online.org/dotherightthing/day5.html.

'Effects of Fast, Cold Front, Set in on Day 2 of the Freedom Fast'. 15 March 2018. [Blog] CIW. https://ciw-online.org/blog/2018/03/freedom-fast-day-2/.

Estabrook, Barry. 2018. *Tomatoland: From Harvest of Shame to Harvest of Hope.* 3rd edition. Kansas City, MO: Andrews McNeel Publishing.

'Fair Food Program Education Travels to Georgia, South Carolina, North Carolina'. 26 June 2017. [Blog] CIW. https://ciw-online.org/blog/2017/06/photos-ffp-education/.

'Fair Food Program Expansion Marches On, Even in These Difficult Times'. 31 July 2020. [Blog] CIW. http://ciw-online.org/blog/2020/07/fair-food-program-expansion-marches-on-even-in-these-difficult-times/.

'Fair Food Season 7, 2018 Update'. CIW. https://www.fairfoodprogram.org/wp-content/uploads/2019/10/Fair-Food-Program-2018-SOTP-Update-Final.pdf.

'Farmworkers to Bike from Immokalee, Carry Iconic Blood-Stained Shirt of Beaten Farmworker to Trader Joe's'. 10 February 2011. [Blog] CIW. https://ciw-online.org/blog/2011/02/feb10_press_release/.

'Farmworker Women Launch "Harvest Without Violence" Campaign in Wendy's Supply Chain!'. 27 September 2017 [Blog] CIW. https://ciw-online.org/blog/2017/09/harvest-without-violence.

Flynn, Alex. 2013. 'Mística, Myself and I: Beyond Cultural Politics in Brazil's Landless Workers' Movement'. *Critique of Anthropology* 33.2: 168–192.

Food Chains: The Revolution in American Fields. 2014. [DVD] United States. Dir. Sanjay Rawal.

Friere, Paulo. 1989a. *Pedagogy of the Oppressed.* Trans. Myra Bergman Ramos. New York: Continuum.

Freire, Paulo with Antonio Faundez. 1989b. *Learning to Question: A Pedagogy of Liberation.* New York: Continuum.

Freire, Paulo. 1994. *Pedagogy of Hope: Reliving 'Pedagogy of the Oppressed'.* Trans. Robert R. Barr. New York: Continuum.

'"Future of Work" Roundtable Spotlights Fair Food Program, Worker-driven Social Responsibility Model!'. 17 October 2018. [Blog] CIW. https://ciw-online.org/blog/2018/10/ffp-wsr-open-democracy-round-table/.

Giagoni, Silvia. 2011. *Fields of Resistance: The Struggle of Florida's Farmworkers for Justice.* Chicago, IL: Haymarket Books.

Gouge, Melissa C. 2016. 'Human Rights in Play, Transnational Solidarity at Work: Creative Playfulness and Subversive Storytelling Among the Coalition of Immokalee Workers'. *Critical Sociology* 42.6: 861–875.

Greenhouse, Steven. 8 November 2018. '"Some Kids Are Not Orphans Because of This": How Unions Are Keeping Workers Safe'. *The Guardian.* https://www.theguardian.com/us-news/2017/nov/08/unions-workers-safety-codes-of-conduct-florida-bangladesh.

Haedicke, Susan. 2020. 'Coalition of Immokalee Workers: Farmworker-Led Popular Education and Performance'. *RIDE: The Journal of Applied Theatre and Performance* 25.4: 576–580. https://doi.org/10.1080/13569783.2020.1804347.

Hammond, John L. 2014. 'Mística, Meaning and Popular Education in the Brazilian Landless Workers Movement'. *Interface: A Journal For and About Social Movements* 6.1: 372–391.

Hassanein, Neva. 2008. 'Locating Food Democracy: Theoretical and Practical Ingredients'. *Journal of Hunger and Environmental Nutrition* 3.2–3: 286–308.

'In Case You Missed It, Wendy's Boycott Goes Global with Epic May Day Virtual Concert for Farmworker Justice'. 7 May 2021. [Blog] CIW. https://ciw-online.org/blog/2021/05/in-case-you-missed-it-wendys-boycott-goes-global-with-epic-may-day-virtual-concert-for-farmworker-justice/.

'Join the Return to Human Rights Tour!'. 22 March 2017. [Blog] CIW. https://ciw-online.org/blog/2017/03/join-the-return-to-human-rights-tour/.

Kester, Grant H. 2004. *Conversation Pieces: Community + Communication in Modern Art*. Los Angeles: University of California Press.

Kester, Grant H. 2011. *The One and the Many: Contemporary Collaborative Art in a Global Context*. Durham and London: Duke University Press.

Klein, Shana. 2018. 'Lady Liberty with a Tomato: A Dialogue on Art and Activism with Kat Rodriguez and The Coalition of Immokalee Farmworkers'. *Public Art Dialogue* 8.1: 98–113.

'Largest Cut Flower Farm on US East Coast joins the Fair Food Program'. 17 August 2020. [Blog] CIW. https://ciw-online.org/blog/2020/08/largest-cut-flower-farm-on-us-east-coast-joins-the-fair-food-program/.

'Latest Slavery Indictments Expose Exploitative Nature of H-2A "Guestworker" Program'. 5 October 2021. [Blog] CIW. https://ciw-online.org/blog/2021/10/latest-slavery-indictments-expose-exploitive-nature-of-h-2a-guestworker-program/?emci=46f92374-db25-ec11-981f-501ac510a405&emdi=8a06d1d3-e225-ec11-981f-501ac510a405&ceid=13379203.

Lee, Wesley. 24 July 2018. 'Farmworkers Target Wendy's Executive in NYC to Demand Better Working Conditions'. *Liberation*. https://www.liberationnews.org/73208-2/.

'A Look Back at the March for Rights, Respect and Fair Food'. 29 March 2013. [Blog] CIW video. https://ciw-online.org/blog/2013/03/video-a-look-back-at-the-march-for-rights-respect-and-fair-food/.

'Major Action Announced: "4 for Fair Food Tour," March 2–14, 2019, Farmworkers and Students to Take the Boot the Braids Campaign to the Next Level!'. 10 December 2018. [Blog] CIW. https://ciw-online.org/blog/2018/12/4-for-fair-food-tour/.

'March for Rights, Respect and Fair Food: Day 15 Photo Report'. 2013. CIW. http://www.ciw-online.org/march/updates/march_17.html?fbclid=IwAR3Ht3dDDV6vcAIeD_UkwN5vZdDBmOXxb3b2T8Kt9vkYrAgBNlcZf3pWBnQ or http://www.ciw-online.org/march/updates/march_17.html.

'Marching with Liberty'. The National Museum of American History. Smithsonian. http://americanhistory.si.edu/many-voices-one-nation/new-americans-continuing-debates-1965%E2%80%932000/marching-liberty.

Marquis, Susan L. 2017. *I Am Not a Tractor! How Florida Farmworkers Took on the Fast Food Giants and Won*. Ithaca and London: Cornell University Press.

'Massive Display of Support by Shareholders for Transparency, Real Human Rights Comes Following Efforts by Company to Block Vote'. 22 May 2021. [Blog]. https://ciw-online.org/.

McMillen, Tracie. 2015/2016. 'Labor Gains'. *Modern Farmer*. http://ciw-online.org/wp-content/uploads/Modern-Farmer-FFP-Article.pdf.

Minkoff-Zern, Laura-Anne. 2014. 'Challenging the Agrarian Imaginary: Farmworker-Led Food Movements and the Potential for Farm Labor Justice'. *Human Geography* 7.1: 85–99.

'Modern Farmer Article Traces History of Fair Food Movement'. 18 December 2015. [Blog] CIW. http://ciw-online.org/blog/2015/12/modern-farmer-article/.

MSI Integrity. 2020. 'Not Fit for Purpose: The Grand Experiment of Multi-Stakeholder Initiatives in Corporate Accountability, Human Rights and Global Governance'. https://www.msi-integrity.org/not-fit-for-purpose/.

'National Farm Worker Ministry Board Leaders Visit Immokalee, Talk Wendy's boycott, Fair Food Program'. 18 February 2020. [Blog] CIW. https://ciw-online.org/blog/2020/02/nfwm-visits-immokalee/.

Nelson, Hilde Lindemann. 1995. 'Resistance and Insubordination'. *Hypatia* 10.2. 23–40.

Nelson, Hilde Lindemann. 2001. *Damaged Identities, Narrative Repairs*. Ithaca, NY: Cornell University Press.

'New Popular Education Drawings Make Their Debut at Worker-to-Worker Sessions of Fair Food Farms Along the East Coast!'. 7 August 2018. [Blog] CIW. https://ciw-online.org/blog/2018/08/northern-education-drawings/.

'New Study Confirms: All Social Responsibility Programs NOT Created Equal'. 21 July 2020. [Blog] CIW. https://ciw-online.org/blog/2020/07/msi-integrity-study/.

'"Not 1996 Anymore…" Worker Beaten at Packing House Near Immokalee'. 15 March 2012. [Blog] CIW. http://ciw-online.org/blog/2012/03/not_1996_anymore/.

'Off Broadway…in Immokalee!'. 9 February 2018. [Blog] CIW video. http://ciw-online.org/blog/2017/02/off-broadway-in-immokalee/.

'One of the Great Human Success Stories of our Day: The Fair Food Program'. 2014. Coalition of Immokalee Workers Information Video. [YouTube]. 2014. https://www.youtube.com/watch?v=BDrOoNGVnJY.

One Penny More. 2010. [YouTube] CIW video. https://www.youtube.com/watch?v=VURs-rsi_KQ&t=4s.

'Over 80 Students and Young People Gather in Immokalee for #2018Encuentro, Launch Plans for Wendy's Boycott Actions in over 30 Cities Nationwide!'. 28 September 2018. http://www.sfalliance.org/news/2018/9/27/over-80-students-and-young-people-gather-in-immokalee-for-2018encuentro-launch-plans-for-wendys-boycott-actions-in-over-30-cities-nationwide.

Paolucci, Paul and Margaret Richardson. 2006. 'Sociology of Humor and a Critical Dramaturgy'. *Symbolic Interaction* 29.3: 331–348.

'Part One: CIW's Women's Group Harvest Without Violence Tour Arrives in Columbus!'. 25 October 2017. [Blog] CIW. https://ciw-online.org/blog/2017/10/part-one-ciw-womens-group-harvest-without-violence-tour-arrives-in-columbus/.

'Photo Report: Hundreds of Students Across the U.S. Demand That Wendy's Leaders Adopt the Unparalleled Human Rights Protections of the Fair Food Program!'. 31 October 2018. Student/Farmworker Alliance. http://www.sfalliance.org/news/2018/10/31/action-report-pulling-back-the-curtain-on-wendys.

'Photo Report, Part 2: "Boot the Braids" Campaign Ramps Up Pressure Against University Leaders' Dishonourable Stance on Wendy's Human Rights Record!'. 4 November 2018. Student/Farmworker Alliance. http://www.sfalliance.org/news/2018/11/4/part-two-boot-the-braids-campaign-ramps-up-pressure-against-university-leaders-dishonorable-stance-on-wendys-human-rights-record.

Polletta, Francesca. 2006. *It Was Like a Fever: Storytelling in Protest and Politics*. Chicago: University of Chicago Press.

Polletta, Francesca. 2008. 'Storytelling in Politics'. *Contexts* 7.4: 26–31.

Polletta, Francesca. 2009. 'Storytelling in Social Movements'. In *Culture, Social Movements, and Protest*. 1st ed. Ed. Hank Johnston. Farnham, Surrey and Burlington, VT: Ashgate. 33–53.

Polletta, Francesca. 2015. 'Characters in Politics'. *Storytelling, Self, Society* 11.1: 34–35.

Polletta, Francesca and Jessica Callahan. 2017. 'Deep Stories, Nostalgic Narratives, and Fake News: Storytelling in the Trump Era'. *American Journal of Cultural Sociology* 5.3: 392–408.

Polletta, Francesca and Pang Ching Bobby Chen. 2012. 'Narrative and Social Movements'. In *The Oxford Handbook of Cultural Sociology*. Eds. Jeffrey C. Alexander, Ronald N. Jacobs and Phillip Smith. Oxford: Oxford University Press. 487–506.

Polletta, Francesca, Pang Ching Bobby Chen, B.G. Gardner, and A. Motes. 2011. 'The Sociology of Storytelling'. *Annual Review of Sociology* 37.1: 109–130.

'Post a Video for "Wendy's the Grinch" Action Today'. 9 December 2016. [Blog] CIW. https://ciw-online.org/blog/2016/12/action-alert-post-a-video-for-wendys-the-grinch-action-today/.

'Publix: Radicals in the Heartland?' 8 February 2021. [Blog] CIW. https://ciw-online.org/blog/2021/02/publix-radicals-in-the-heartland/.

Rancière, Jacques. 2006. 'Problems and Transformations in Critical Art'. In *Participation*. 1st ed. Ed. Claire Bishop. London: Whitechapel. 83–93.

Rape in the Fields. 2013. [TV Programme] PBS Frontline. https://www.pbs.org/wgbh/frontline/film/rape-in-the-fields/.

Re: Structure Lab. 2021. 'Due Diligence and Transparency'. https://static1.squarespace.com/static/6055c0601c885456ba8c962a/t/60660e41b634ac7381898670/1617301058609/ReStructureLab_DueDiligenceAndTransparencyLegislation_April2021.pdf.

Renz, Katie. 2005. 'Power People: An Interview with David Solnit'. *Mother Jones*. https://www.motherjones.com/politics/2005/03/people-power-interview-david-solnit/.

Reyes, Gerardo. 2018. 'Gerardo Reyes Chavez: What Is WSR?' [YouTube] CIW video. https://www.youtube.com/watch?v=yetarZVHrHg.

Rodriguez, Nely. 5 October 2018. *Personal Interview*.

'Scores of Students, Youth Gather in Immokalee for #2018Encuentro, Launch Plans for Nationwide Wendy's Boycott Actions!'. 2018. https://ciw-online.org/blog/2018/10/encuentro-report/.

'Social Media Chatter Surges Around the March for Rights, Respect and Fair Food!'. 1 March 2013. [Blog] CIW. https://ciw-online.org/blog/2013/03/social-media-chatter-surges-around-the-march-for-rights-respect-and-fair-food/.

Solnit, David. 2012. 'Excerpts from David Solnit's Comments During Part 1 of the Series, "Revolutionary Nonviolence?"'. [YouTube] https://www.youtube.com/watch?v=uxFj4xP4j4A.

Solnit, David. 2014a. 'Arts Organizing & Strategy for Revolution'. Part 1: Cantastorias, Making Visuals, Why Organize?'. [YouTube]. https://www.youtube.com/watch?v=EfEKcXHaE94.

Solnit, David. 2014b. 'Arts Organizing & Strategy for Revolution'. Part 2. [YouTube]. https://www.youtube.com/watch?v=cJJ_u0KOpBE.

Solnit, David. 2016. 'Beyond Resolution. *Artists Now! David Solnit*'. [YouTube]. https://www.youtube.com/watch?v=kZRRkEu3r3E.

'Starting today (Friday 11/10), the Harvest Without Violence Campaign hits New York City!'. 10 November 2017. [Blog] CIW. https://ciw-online.org/blog/2017/11/harvest-without-violence-nyc/.

The Story Behind Your Food. 2018. [YouTube] CIW video. https://www.youtube.com/watch?v=Ny9YaJczqjY.
Student Farmworker Alliance. http://www.sfalliance.org/.
'Support for the Fast, Campaign, Wide, Deep'. 16 March 2012. [Blog] CIW. https://ciw-online.org/blog/2012/03/fast_support_post/.
'Take to Social Media: #BOYCOTTWENDYS'. 2020. http://www.boycott-wendys.org/take-action/.
'Take to Social Media to Urge Wendy's Top Investors to Leverage Their Power to Bring the Fast Food Giant into the Fair Food Program!'. 9 March 2020. [Blog] CIW. http://ciw-online.org/blog/2020/03/follow-the-money-social-media/.
A Tale of Two Roadside Farm Stands. 2 February 2017. [Blog] CIW. https://ciw-online.org/blog/2017/02/a-tale-of-two-roadside-farm-stands/.
Taylor, Diana. 2003. *The Archive and the Repertoire: Performing Cultural Memory in the Americas*. Durham and London: Duke University Press.
'Theater, Reflection Make for a Festive Evening in Immokalee!'. 28 February 2013. [Blog] CIW. https://ciw-online.org/blog/2013/02/theater-reflection-make-for-a-festive-evening/.
Thomas, Gillian. 2018. '#MeToo Can't Fix the Workplace. Here's an Option that Can'. *The Washington Post*. https://www.washingtonpost.com/outlook/metoo-hasnt-fixed-the-workplace-heres-a-playbook-that-can/2018/11/29/bb1de86e-e762-11e8-b8dc-66cca409c180_story.html?noredirect=on&utm_term=.aed041efb823.
Tiziou, J.J. 2017. 'CIW & Fair Food'. [Photographic album]. https://www.jjtiziou.net/series/ciw-fair-food.
'A Truly GRAND Finale!'. 18 March 2019. [Blog] CIW. https://ciw-online.org/blog/2019/03/a-truly-grand-finale/.
Valdez, Luis. 1994. *Luis Valdez—Early Works: Actos, Bernabé and Pensamiento Serpentino*. 2nd printing. Kindle edition. Houston, TX: Arte Público Press.
'VIDEO Action: Boycott the Fast Food Grinch'. 2016. https://www.facebook.com/events/179524385851155/?active_tab=discussion.
'With the Coalition of Immokalee Workers—Sets for a Mística Theatre at the March for Rights, Respect and Fair Food'. 2013. https://monacaron.com/CIW-mistica.
'With the Coalition of Immokalee Workers—Visuals for Mass Action by Florida's Farmworkers'. 2011. https://monacaron.com/artivism/coalition-immokalee-workers-visuals-mass-action-floridas-farmworkers.
'Worker-driven Social Responsibility—Beyond the Headlines'. 2020. [Video] Worker-driven Social Responsibility Network. https://wsr-network.org/video/.
WSR: Worker-driven Social Responsibility Network. 2020. 'What Is WSR?' https://wsr-network.org/what-is-wsr/.

'"Yon Sel Ko": Timely Wisdom from the Haitian Peasant Movement at the Root of the CIW'. 19 March 2020. [Blog] CIW. https://ciw-online.org/blog/2020/03/yon-sel-ko/.

Zwerdling, Daniel. 2005. 'Fast-Food Deal a Big Win for Small Migrants' Group'. NPR (National Public Radio). *All Things Considered*. https://www.npr.org/templates/story/story.php?storyId=4706271.

CHAPTER 7

Conclusion: Sowing Seeds and Cultivating Conversations for Future Harvests

Performing complex farmscapes could not be more important to our world in its present damaged state. Artists, like the ones in this book, address the often devastating problems in today's global agri-food system in layman's words and life stories, making them comprehensible and compelling. Their art plays an important role in altering cultural constructions of farmlands so that they more accurately reflect current agricultural realities and dispel idyllic fantasies of the countryside. The performed and performing farmscapes galvanise audiences to be concerned about the urgency of developing a just and sustainable food system—a food democracy—to protect the planet and all living organisms on it. And their performativity engages the public and so contributes to a transformation from disempowered food *consumer* at the end of the food chain to empowered food *citizen* helping to shape an ethical food system. The artists in the case studies apply concepts and strategies of food citizenship in their works even though they do not mention those words by name. In so doing, the book argues, they not only inform their audiences about these vital issues, but also capture their attention through dialogue and participation and facilitate radical thinking about what an improved agri-food system could be. This new knowledge can convince audiences to respond to the varied crises caused by the food system today by altering their food behaviours so that they promote justice and sustainability: anything from joining CSAs or changing what they buy

© The Author(s), under exclusive license to Springer Nature Switzerland AG 2021

S. C. Haedicke, *Performing Farmscapes*, Performing Landscapes, https://doi.org/10.1007/978-3-030-82434-1_7

to reflect their support for humane treatment of animals, fair working conditions for farmworkers and stewardship of the environment to taking part in marches and boycotts, establishing networks with the agricultural community and participating in policy development. This new behaviour, in turn, changes food production, distribution and consumption practices locally, regionally, nationally and globally as the many writers on food democracy[1] and the case studies themselves confirm. A well-informed and engaged population is key to ameliorating an unethical agri-food system even if the changes are slow in coming and sometimes temporary. Food citizens believe in the power of ordinary people to shape the future of food.

While educating the public about our broken food systems and engaging them in finding solutions are about encouraging food ethics, it is also essential to our survival. The Food and Agriculture Organization of the United Nations (FAO) identifies food insecurity[2] as one of the most pressing global issues and estimates that two billion people in the world are food insecure and that 690 million people (or approximately one in ten people worldwide) actually go hungry. Although in decline for decades, the number of the food insecure began to rise in 2015 increasing by 60 million by 2019. As this book goes to press, preliminary estimates indicate that the COVID-19 pandemic has caused that number to rise by an additional 132 million in 2020 (FAO 2020: 4, 11–12).[3] In spite of these shocking statistics, current global food production can feed the entire population (FAO 2020: 8), but capitalist interests, chronic poverty[4] and geo-political conflicts make food distribution inequitable (FAO 2020: 6, 14; Action Against Hunger 2021) with the largest numbers of the food insecure in Africa. Climate change worsens food insecurity as extreme

[1] See Civil Eats (2021), Food Ethics Council (2021), and Union of Concerned Scientists (n.d.) as well as authors cited in Chapter 1.

[2] The food insecure do not have regular access what are considered the four pillars of food security: availability, access, stability and utilization (nutrition). See, for example, FAO (2020); Martin-Shields and Stojetz (2019); Food Security Information Network (2020) and McGovern and Hayes (2021).

[3] See also Civil Eats (2021) and United Nations (n.d.).

[4] The Food Ethics Council claims that in terms of proportion of income, the poor in the UK spend twice as much on food as the rich, according to the UK Office of National Statistics. The Report argues that cheap food and food banks just exacerbate the problem (Food Ethics Council 2017: 37).

weather conditions destroy crops and kill livestock. Paradoxically, agriculture is one of the major contributors to climate change. Current agricultural practices cause 21–37% of greenhouse gas emissions (FAO 2020: 31) and damage the earth's soil, water, biodiversity and more. While high yield crops grown with chemical fertilisers and pesticides certainly have made low-cost food more available, it is often less nutritious, and it is hard to overlook the high profits to growers and chemical companies from this agricultural practice that devastates the land and its inhabitants and creates huge wealth discrepancies. Not surprisingly, food insecurity plays a major role in adversely affecting human health with disorders ranging from malnutrition to obesity.[5] FAO claims that healthcare costs are one of the hidden costs of the current food system and could be reduced by 97% with the availability of a healthy diet for the world's population (FAO 2020: 29). In addition, recent scholarship has found that food insecurity is a leading cause of violent conflict that can escalate to war.[6] Food insecurity is clearly the root of a myriad of social, political, economic and environmental crises. Understanding its sources and effects and devising an ethical food system that supplies adequate, affordable and nutritious food for all are key to our survival.[7]

In *Performing Farmscapes*, I have argued that *farmscapes* are not just agricultural lands that produce food for the world's population or iconic rural landscapes, although the concept of 'farmscape' certainly includes both of those definitions. Farmscapes are gatherings of spatial–temporal narratives told in the words and experiences of farmers, in the rhythms

[5] In addition, the World Health Organization (WHO) has found that almost one in ten people around the world suffers from mild to severe food poisoning from contaminated food each year (WHO 2015: n.p.). See also WHO (n.d.).

[6] See, for example, Bruck and d'Errico (2019) and Martin-Shields and Stojetz (2019) in a special issue of *World Development* 119 that looks at the impact of food insecurity on initiating or escalating conflicts rather than conflicts causing food insecurity, but there are many other studies.

[7] Such a food system would, for example, fight against environmental damage caused by agriculture and work towards providing safe and fair working conditions for farm labourers and humane treatment of livestock, securing governmental support for sustainable agricultural practices and achieving gender and race equality in farming that ensures equal access to land and equipment. Action Against Hunger claims that if women farmers' access to resources was equivalent to men's, the number of hungry people would decrease by 17% with the increase in crop yields.

and repetitions of seasonal farming practices associated with a particular place and in the multispecies encounters with each other and with the land. The book's pairing, or 'duet', of farmscapes and performing makes these narratives visible and clarifies their complex web of storylines, pathways, experiences, encounters, emotions, memories and echoes by creating patterns and rhythms. The duet also suggests a potential for agency for farmlands to participate in the conversations through physical responses to human and nonhuman interventions recognisable in the traces left in the land by living creatures and natural forces alike and visible in seemingly out-of-place landscape features, like the wolf tree. *Performing Farmscapes* looks at how contemporary artists use performance and performativities to collaborate with farmscapes to remember past experiences, engage with present-day realities and realise future possibilities that can lead to an ethical agri-food system. The book introduces a theoretical framework to interpret farmscapes through three distinct (although simultaneously overlapping) modalities that offer ways of knowing and experiencing them. The first modality focuses on the significance of the local for an understanding of farmscapes. It foregrounds the importance of insider knowledge of the specificities of a particular farmscape in terms of topography, climate and multispecies' behaviour, practices and attitudes. The second modality looks at farmscapes through a lens of agricultural work and argues that not seeing the hard work involved in farming leads to a perception of farmlands as pastoral landscapes and often to exploitation of the invisible farmworkers. The final modality explores the inseparability of nature and culture in storied farmscapes and links survival of all living creatures to care for the earth. The underlying assumption of analysing these performing landscapes through local, working and naturalcultural modalities is that each one addresses an interpretive approach to farmscapes that can influence attitudes and behaviour, but it is important to note that in a farmscape, they are intertwined.

The core argument of the book is that the performance-based work in the case studies informs and involves the public in agricultural issues and so contributes to the transformation of passive and uninformed food *consumers* whose identity is linked to what they buy into proactive and educated food *citizens* whose identity is linked to what they do (Food Ethics Council 2014: 1, 17). That shift in identity can inspire attitudinal and behavioural shifts as concern for the common good replaces self-interest. While the performances discussed in *Performing*

Farmscapes certainly do communicate complex scientific research about food production, distribution and consumption in more accessible and engaging ways than an academic paper or a policy report, they are not just agricultural information. The artists translate esoteric data into life stories and create spaces to experiment with and 'rehearse' new ideas and behaviours through discussions, workshops and collaborative art-making to encourage audiences to reflect on an ethical agri-food system and collaborate on making it a reality. As a result, these spectator-participants can move beyond a sense of powerlessness to an awareness of their agency to effect social change. After setting out the core argument in the Introduction (Chapter 1), the book traces a path from food consumerism to food citizenship by organising the case studies to reflect a trajectory that moves from projects that foreground knowledge acquisition to ones that emphasise participatory social engagement, to a final one that has succeeded in changing aspects the migrant farmscape in the eastern United States. The projects discussed in the two chapters following the Introduction focus on varied performance-based creative practices to share knowledge about local, working, naturalcultural farmscapes, to interest audiences in crises facing food production and to plant the idea that the food system can be changed. These projects focus on insider knowledge provided by the farmers' words and agricultural practices (Chapter 2) and by traces on the land that narrate stories of the farms' pasts (Chapter 3). The next two chapters look at projects that foreground conversations and participation to increase knowledge about food production and distribution. These projects initiate a process of politicisation by creating spaces for discussion, 'rehearsals' for behaviours of citizenship and collaborative art-making (Chapter 4) or, alternatively, by shocking audiences with agricultural narratives, often about exploitation and abuse of land and labour, to mobilise them into action (Chapter 5). The final chapter looks at the efficacy of food citizenship in the Coalition of Immokalee Workers' Worker-driven Social Responsibility (WSR) model that has achieved verifiable changes in food production on monocrop industrialised farms. Here, the pairing of farmworker-led activism and protest art in partnership with students, religious groups and other food citizens achieves legally binding results (Chapter 6).

In 1971, artist Robert Smithson, concerned about the ecological damage caused by the mining industry, made an appeal to artists to intervene: 'Art can become a resource, that mediates between the ecologist and the industrialist. Ecology and industry are not one-way streets, rather they should be crossroads. Art can help provide the needed dialectic between them' (Smithson 2019: 208). Smithson's call to miners and artists to work together, 'conscious of themselves as natural agents' (Smithson 2019: 208), can apply to farmers and artists today. Donna Haraway echoes Smithson's call for collaboration across disciplines as she asserts the efficacy of creative art-science worldings as sympoietic practices (Haraway 2016: 67) to *make-with* across species.[8] Her worlding story emphasises the urgency of 'staying with the trouble', of being fully present, aware, 'response-able' and 'becoming-with' or 'making-with' all the 'critters' in an interlaced 'naturalcultural' world (terms used repeatedly in Haraway 2016). Staying with the trouble demands a persistent focus on current problems rather than a fixation on past or future disasters, and *response-ability* necessitates collaboration and so is our best hope to heal our damaged world. The case studies in this book provide examples of art-agriculture worldings, ongoing processes of sympoiesis, of *making-with*, that encourage *response-ability* as audiences participate in the art. In this practice of 'commoning' that advocates sharing resources, ideas and experiences over exploiting each other or the planet, success is measured by the well-being of the community as a whole, not individual gain. Commoning is a key aspect of food democracy and seen as an essential step in enabling the planet and its life forms to survive.[9] (Fig. 7.1).

Performing Farmscapes marks just a starting point in an inquiry into these creative worldings and suggests other ways to approach art-agriculture projects from other places around the world certainly, but also through different lenses. The range of performance-based/farm-based projects explored in the book argues against the existence of a specific 'farmscape performance' genre and instead encourages open-ended thinking that proposes new ways of engaging with art and agriculture together. Lippard claims that 'art about agriculture…will not be fully effective until it goes underground, until it is integrated into and

[8] 'Unsettled Nature: Artists Reflect on the Age of the Human', an exhibition at the Smithsonian Museum of Natural History that opened in June 2021, turns to art-science worldings to explore the detrimental impact of humans on the earth. The museum's curator Scott Wing said they turned to art since the scientists' warnings 'don't always get across'. Photographer Dornith Doherty, for example, comments on a crop's or seed's human-caused evolutionary changes in her documentation of potato plant clones or seeds in crop seed banks (Catlin 2021).

[9] This idea is a leitmotif throughout writings on food citizenship, many of which are listed in Chapter 1. See Food Ethics Council for many recent reports and articles.

Fig. 7.1 *Adapting. Evolving. Growing.* Pen and ink drawing by Kristen Oshyn

almost disappears into the local culture and nature itself' (Lippard 2007: 45). While this may be an overstatement, I do think it highlights the need to rethink ways to assess the value of art-agriculture projects by turning away from looking at reviews, audience development, 'aesthetics' and even impact and instead moving towards an understanding of how these current and future projects are integral to politicising people to be proactive in ensuring justice and sustainability in our food system. How, for example, can new art-agriculture worldings contribute to meeting Goal 2: Zero Hunger of the United Nations Sustainable Development Goals, signed by the countries of the United Nations, that sets 2030 as the date for ensuring access to safe, nutritious and sufficient food for all people all year round? (United Nations n.d. and FAO 2020).[10] How can art-agriculture projects convince governments and corporations to fund sustainable agricultural practices even if they reduce financial profits? How can they make unsafe working conditions and unequal access to

[10] See also McGovern and Hayes (2021).

productive resources unpalatable to farming and non-farming communities? How can they make food insecurity for *anyone* an anathema to all people? The duet of performing and farmscapes can suggest varied ways to become food citizens working together with other species and materialities to find tools for navigating our way through the socio-political, economic and ecological challenges to an ethical agri-food system that we face as we work our way through the twenty-first century. The performance-based case studies and the drawings throughout the book have constantly advanced my ideas or overturned my ways of thinking. They suggest original, creative paths forward and insist that we cannot turn away. We must 'stay with the trouble' that the artists reveal. So I give the art the final 'word' (Fig. 7.2).

Fig. 7.2 *Coalescence*. Pen and ink drawing by Kristen Oshyn

Works Cited

Action Against Hunger. 2021. 'Food Security and Livelihoods'. https://www.actionagainsthunger.org/impact/food-security-livelihoods.

Bruck, Tilman and Marco d'Errico. 2019. 'Food Security and Violent Conflict: Introduction to the Special Issue'. *World Development* 119: 145–149.

Catlin, Roger. 2021. 'The Sad Truths behind these Unsettling Works of Art'. *Smithsonian Magazine*. https://www.smithsonianmag.com/smithsonian-institution/unsettling-experience-viewing-natural-disasters-works-art-180977949/.

Civil Eats. 2021. https://civileats.com/.

Food and Agriculture Organization of the United Nations. 2020. *In Brief. The State of Food Security and Nutrition in the World: Transforming Food Systems for Affordable Healthy Diets*. http://www.fao.org/3/ca9699en/ca9699en.pdf.

Food Ethics Council. 2014. 'This is the #citizenshift: A Guide to Understanding and Embracing the Emerging Era of the Citizen'. https://drive.google.com/file/d/1ce2qAxXjgw30KuuZTH9BKTw1xCUGV7Us/view.

Food Ethics Council. 2017. 'Food Citizenship: How Thinking About Ourselves Differently Can Change the Food System'. https://www.foodethicscouncil.org/app/uploads/FoodCitizenship2017_web.pdf.

Food Ethics Council. 2021. https://www.foodethicscouncil.org/.

Food Security Information Network. 2020 Global Report of Food Crises. 2020. https://docs.wfp.org/api/documents/WFP-0000114546/download/?_ga=2.16074431.840132338.1623705097-305804507.1623705097.

Haraway, Donna. 2016. *Staying with the Trouble: Making Kin in the Chthulucene*. Durham, NC and London: Duke University Press.

Ingold, Tim. 2011. *Being Alive: Essays on Movement, Knowledge and Description*. Abingdon, UK: Taylor & Francis Group.

Lippard, Lucy. 2007. 'A Frame for Farming'. In *Focus on Farmers: Art and Hill Farming*. Ed. Aune Head Arts. Bayley Greene, Wellington, Somerset, UK: Halsgrove. 38–45.

Martin-Shields, Charles and Wolfgang Stojetz. 2019. 'Food Security and Conflict. Empirical Challenges and Future Opportunities for Research and Policy Making on Food Security and Conflict'. *World Development* 119: 150–164.

McGovern, James and Jahana Hayes. 2021. 'Hunger is a Political Issue. We Can Work to End it'. Civil Eats. https://civileats.com/2021/05/26/op-ed-hunger-is-a-political-decision-we-can-work-to-end-it/.

Smithson, Robert. 2019. 'Untitled'. In *The Rural*. Ed. MyVillages. London: White Chapel Gallery and Cambridge, MA: The MIT Press. 208.

Union of Concerned Scientists. n.d. 'Food Justice'. https://www.ucsusa.org/food/food-justice.

United Nations. n.d. 'Sustainable Development Goals. Goal 2: Zero Hunger'. https://www.un.org/sustainabledevelopment/hunger/.
World Health Organization. n.d. https://www.who.int/health-topics/food-safety/.
World Health Organization. 2015. 'WHO Estimates the Global Burden of Foodborne Diseases'. https://apps.who.int/iris/bitstream/handle/10665/199350/9789241565165_eng.pdf?sequence=1.

Bibliography

Action Against Hunger. 2021. 'Food Security and Livelihoods'. https://www.actionagainsthunger.org/impact/food-security-livelihoods.

'Advantages and Disadvantages of Intensive Farming'. 2018. *Farm Management*. http://www.farmmanagement.pro/advantages-and-disadvantages-of-intensive-farming/.

AgArts. 2021. https://www.agarts.org/.

Allegue, Ludivine, Simon Jones, Baz Kershaw, and Angela Piccini, eds. 2009. *Practice-as-Research in Performance and Screen*. Basingstoke: Palgrave Macmillan.

Alliance for Fair Food. http://www.allianceforfairfood.org.

'The Alliance for Fair Food: Uniting for a New Day in the Fields' 2018. [YouTube video] https://www.youtube.com/watch?v=SYfKFmGNghQ&t=178s.

Angel, Jen. 2008. 'David Solnit and the Arts of Change'. www.joaap.org/webonly/solnit_angel.htm.

Ansley, Fran and John Gaventa. 1997. 'Researching for Democracy and Democratizing Research'. *Change: The Magazine of Higher Learning* 29.1: 46–53.

Anderson, Ben and Paul Harrison. 2010. 'The Promise of Non-Representational Theories'. In *Taking-Place: Non-Representational Theories and Geography*. Eds. Ben Anderson and Paul Harrison. Farnham, Surrey and Burlington, VT: Ashgate. 1–35.

Arlander, Annette, Bruce Barton, Melanie Dreyer-Lude, and Ben Spatz, eds. 2018. *Performance as Research: Knowledge, Methods, Impact*. London and New York: Routledge.

© The Editor(s) (if applicable) and The Author(s) 2021

S. C. Haedicke, *Performing Farmscapes*, Performing Landscapes,
https://doi.org/10.1007/978-3-030-82434-1

Austin, J.L. 1976 [1955]. *How To do Things with Words: the William James Lectures Delivered at Harvard University in 1955*. 2nd ed. Oxford: Oxford University Press.

Bacon, David. 2002. 'No Quiero Taco Bell! David Bacon interviews Lucas Benitez of the Coalition of Immokalee Workers'. *In These Times*. https://inthesetimes.com/article/no-quiero-taco-bell.

Bagby, Beth and Luis Valdez. 1967. 'El Teatro Campesino Interviews with Luis Valdez'. *The Tulane Drama Review* 11.4: 70–80.

Balmford, Andrew, et.al. 2008. 'Migrant Tomato Workers Face Chronic Abuses'. Testimony before the US Senate Committee on Health, Education, Labour and Pensions. Southern Poverty Law Center. https://www.splcenter.org/news/2008/04/15/migrant-tomato-workers-face-chronic-abuses.

Baym, Geoffrey. 2010. *From Cronkite to Colbert: The Evolution of Broadcast News*. Boulder, CO: Paradigm.

Benediktsson, Karl and Katrín Anna Lund. 2010. *Conversations with Landscape*. Farnham, Surry, England and Burlington, VT: Ashgate.

Bennett, Jane. 2010. *Vibrant Matter: A Political Ecology of Things*. Durham and London: Duke University Press.

Berry, Wendell. 2010 [1990]. *What are People For? Essays*. 2nd ed. Berkeley, CA: Counterpoint.

Beuys, Joseph. 2006. 'I Am Searching for Field Character, 1973'. In *Participation*. Ed. Claire Bishop. London: Whitechapel. 125–126.

Bishop, Claire. 2004. 'Antagonism and Relational Aesthetics'. *October* 110: 51–79.

Bishop, Claire. 2006. 'The Social Turn: Collaboration and Its Discontents'. *Artforum* February: 178–183. https://www.artforum.com/print/200602/the-social-turn-collaboration-and-its-discontents-10274.

Bishop, Claire. 2009. 'The Social Turn: Collaboration and Its Discontents'. In *Rediscovering Aesthetics: Transdisciplinary Voices from Art History, Philosophy, and Art Practice (Cultural Memory in the Present)*. Eds. Francis Halsall, Julia Jansen, and Tony O'Connor. Stanford: Stanford University Press. 238–255.

Bishop, Claire. 2012. *Artificial Hells: Participatory Art and the Politics of Spectatorship*. London and New York: Verso.

Blair, Elizabeth. 2014. *All Things Considered: Revisiting The Stories Told in Murrow's 'Harvest of Shame'*. [Radio programme] NPR. Transcript. https://www.npr.org/2014/11/28/367244290/revisiting-the-stories-told-in-murrows-harvest-of-shame.

Böhm, Steffen, Marja Spierenburg, and Tim Lang. 2020. 'Fruits of Our Labour: Work and Organisation in the Global Food System'. *Organisation* 27.2: 195–212.

Borda, Sylvia Grace. 2015. *Farm Tableaux Finland*. http://www.sylviagborda.com/.

Borda, Sylvia Grace in collaboration with John M. Lynch. 2015. *Farm Tableaux Finland Google Street View*. [YouTube] https://www.youtube.com/watch?v=wPqndBnCcxM.

Bowe, John. 2003. 'Nobodies Annals of Labor'. *The New Yorker* 79.9: 106–133.

Braham, Emily. 2016. 'Milking Art to Improve Food Ethics'. *Positive News*. https://www.positive.news/environment/milking-art-to-improve-food-ethics/.

Brandes, Stanley. 1998. 'Iconography in Mexico's Day of the Dead: Origins and Meanings'. *Ethnohistory* 45.2: 181–218.

Broyes-Gonzalez, Yolanda. 1994. *El Teatro Campesino: Theatre in the Chicano Movement*. Austin: University of Texas Press.

Broyes-Gonzalez, Yolanda. 2006. 'Re-constructing Collective Dynamics: El Teatro Campesino from a Twenty-first Perspective'. In *Restaging the Sixties: Radical Theaters and Their Legacies*. Eds. James M. Harding and Cindy Rosenthal. 1st ed. Ann Arbor: University of Michigan Press.

Bruck, Tilman and Marco d'Errico. 2019. 'Food Security and Violent Conflict: Introduction to the Special Issue'. *World Development* 119: 145–149.

Bruzzi, Stella. 2006. *New Documentary*. 2nd ed. London and New York: Routledge.

Bullard, Gabe. 2015. 'How to Update *Harvest of Shame* for the 21st Century'. *National Geographic*. https://www.nationalgeographic.com/people-and-culture/food/the-plate/2015/11/24/how-to-update-harvest-of-shame-for-the-21st-century/.

Canny, Autumn L. 2005. 'Lost in a Loophole: The Fair Labor Standards Act's Exemption of Agricultural Workers from Overtime Compensation Protection'. *Drake Journal of Agricultural Law* 10: 356–386.

Carolan, Michael S. 2008. 'More-than-Representational Knowledge/s of the Countryside: How We Think as Bodies'. *Sociologia Ruralis* 48.4: 408–422.

Caron, Mona. 2011a. 'With the Coalition of Immokalee Workers—Visuals for Mass Action by Florida's Farmworkers'. https://monacaron.com/artivism/coalition-immokalee-workers-visuals-mass-action-floridas-farmworkers or https://monacaron.com/artivism/coalition-immokalee-workers-visuals-mass-action-floridas-farmworkers.

Caron, Mona. 2011b. 'Do the Right Thing'. https://www.facebook.com/media/set/?set=a.1859178880548&type=3.

Caron, Mona. 2013a. 'With the Coalition of Immokalee Workers—Sets for a mística theatre at the March for Rights, Respect and Fair Food'. https://monacaron.com/CIW-mistica.

Caron, Mona. 2013b. 'CIW 200 Mile March across Florida for Rights, Respect and Fair Food'. https://www.facebook.com/mona.caron/media_set?set=a.10200717914091569.1073741831.1273193655&type=3.

Carruth, Allison. 2013. *Global Appetites: American Power and the Literature of Food*. Los Angeles: University of California Press.

Catlin, Roger. 2021. 'The Sad Truths behind these Unsettling Works of Art'. *Smithsonian Magazine*. https://www.smithsonianmag.com/smithsonian-institution/unsettling-experience-viewing-natural-disasters-works-art-180977949/.

'CBS Sunday Morning: Fair Food Program'. 2015. [TV programme] https://www.youtube.com/watch?v=b6s4YAOISns.

Civil Eats. 2021. https://civileats.com/.

Cloke, Paul and Owain Jones. 2004. 'Turning in the Graveyard: Trees and the Hybrid Geographies of Dwelling, Monitoring and Resistance in a Bristol Cemetery'. *Cultural Geographies* 11.3: 313–341.

CNN 'Freedom Project'. 2017. https://edition.cnn.com/2017/05/30/world/ciw-fair-food-program-freedom-project/index.html .

CNN 'Freedom Project—Wendy's Boycott'. 2017. [TV programme] https://www.youtube.com/watch?v=OKm_yZvBOJ0.

Coalition of Immokalee Workers. https://ciw-online.org/.

Coalition of Immokalee Workers. 2000–2021. Blog. CIW. https://ciw-online.org/?s=blog.

'Coalition of Immokalee Workers: La Mixteca—Street Theatre. 2003 Taco Bell Truth Tour'. 2003. [YouTube]. https://www.youtube.com/watch?v=bB1zNasI_Dc.

Bourriaud, Nicolas. 2002. *Relational Aesthetics*. Trans. Simon Pleasance and Fronza Woods. Dijon-Quetigny: Les Presses de reel.

Con Estas Manos/With these Hands. [YouTube] CIW video. https://www.youtube.com/watch?v=ItRPBp8kj8o.

Cook, Ian. 2000. 'Social Sculpture and Connective Aesthetics: Shelley Sacks's *Exchange Values*'. *Ecumeme* 7.3: 337–343.

Cosgrove, Denis. 2004. 'Landscape and Landschaft'. *German Historical Institute Bulletin* 35: 57–71.

Council of Europe (COE). 2000. European Landscape Convention and Reference Documents. https://rm.coe.int/CoERMPublicCommonSearchServices/DisplayDCTMContent?documentId=09000016802f80c6.

Coveney, J. 2014. *Food*. Abingdon, Oxon: Routledge.

Crouch, David. 2010. *Flirting with Space: Journeys and Creativity*. Farnham, UK and Burlington, VT: Ashgate.

Crouch, David. 2013. 'Landscape, Performance and Performativity'. In *The Routledge Companion to Landscape Studies*. Eds. Peter Howard, Ian Thompson, and Emma Waterton. London and New York: Routledge. 119–127.

Cura, Anna and Dan Crossley. 2021. 'Harnessing the Power of Food Citizenship'. Food Ethics Council. https://foodcitizenship.info//wp-content/uploads/2019/10/fec_food_citizenship_report_final.pdf.

Deleuze, Gilles. 1995. *Negotiations: 1972–90*. Trans. Martin Joughin. New York: Columbia University Press.

Denes, Agnes. n.d. *Wheatfield: a Confrontation*. http://www.agnesdenesstudio.com/works7.html.

Descola, Philippe. 2013. *Beyond Nature and Culture*. Trans. Janet Lloyd. Chicago and London: The University of Chicago Press.

Deutsche, Rosalyn. 1996. *Evictions: Art and Spatial Politics*. Cambridge, MA and London: The MIT Press.

Devine, Jenny. 2013. *On Behalf of the Family Farm*. Iowa City, Iowa: University of Iowa Press.

Dove, Rita. 1994. *The Darker Face of the Earth*. London: Oberon Books, Ltd.

'Drama of the Land'. 2015. [Video] Harvest Public Media. https://maryswander.com/media-appearances/.

Dixon, Beth A. 2018. *Food Justice and Narrative Ethics*. London: Bloomsbury.

'Do the Right Thing: Day 5 Update'. 2011. http://www.ciw-online.org/dotherightthing/day5.html.

Dewsbury, John-David. 2000. 'Performativity and the Event: Enacting a Philosophy of Difference'. *Environment and Planning D: Society and Space* 18: 473–496.

Estabrook, Barry. 2018. *Tomatoland: From Harvest of Shame to Harvest of Hope*. 3rd ed. Kansas City, MO: Andrews McNeel Publishing.

European Coordination Via Campesina: Nyéléni Europe. n.d. https://www.eurovia.org/campaign/nyeleni/.

Fahlenbrach, Martin Klimke and Joachim Scharloth, eds. 2016. *Protest Cultures: A Companion*. New York and Oxford: Berghahn.

'Farmworker Justice'. US Department of Labor. 2019. http://www.farmworkerjustice.org/wp-content/uploads/2019/05/NAWS-Data-FactSheet-05-13-2019-final.pdf.

A Field of Wheat: Arts and Agricultural Project. n.d. http://fieldofwheat.co.uk/.

Filewod, Alan. 1987. *Collective Encounters: Documentary Theatre in English Canada*. Toronto, Buffalo and London: University of Toronto Press.

Filewod, Alan. 2008. 'Collective Creation: Process, Politics and Poetics'. In *Collective Creation, Collaboration and Devising*. Ed. Bruce Barton. Toronto: Playwrights Canada Press. 1–13.

Fischer-Lichte, Erika. 2008. *The Transformative Power of Performance: A New Aesthetics*. Trans. Saskya Iris Jain. London and New York: Routledge.

Flynn, Alex. 2013. 'Mística, Myself and I: Beyond Cultural Politics in Brazil's Landless Workers' Movement'. *Critique of Anthropology* 33.2: 168–192.

Food and Agriculture Organization of the United Nations. 2020. *In Brief. The State of Food Security and Nutrition in the World: Transforming Food Systems for Affordable Healthy Diets*. http://www.fao.org/3/ca9699en/ca9699en.pdf.

Food Chains: The Revolution in American Fields. 2014. [DVD] United States. Dir. Sanjay Rawal.

Food Ethics Council. 2021. https://www.foodethicscouncil.org/.

Food Security Information Network. 2020. Global Report of Food Crises. 2020. https://docs.wfp.org/api/documents/WFP-0000114546/download/?_ga=2.16074431.840132338.1623705097-305804507.1623705097.

Fordhall Organic Farm. https://www.fordhallfarm.com/.

Freemantle, Chris. n.d. 'A Field of Wheat: Whose Art?' http://fieldofwheat.co.uk/wp-content/uploads/2016/09/A-Field-of-Wheatcfremantle.pdf.

Friere, Paulo. 1989. *Pedagogy of the Oppressed*. Trans. Myra Bergman Ramos. New York: Continuum.

Freire, Paulo. 1994. *Pedagogy of Hope: Reliving 'Pedagogy of the Oppressed'*. Trans. Robert R. Barr. New York: Continuum.

Freire, Paulo with Antonio Faundez. 1989. *Learning to Question: A Pedagogy of Liberation*. New York: Continuum Publishing Company.

Friches Théâtre Urbain. 2012. *L'Espwar est un temps boisé*. https://www.friches.fr/projets/l-espwar-est-un-temps-boise and http://parcelle343.hautetfort.com/about.html.

Fuchs, Elinor. 2002. 'Reading for Landscape: The Case of American Drama'. In *Land/Scape/Theatre*. Eds. Elinor Fuchs and Una Chaudhuri. Ann Arbor: University of Michigan Press. 30–50.

Gabriel, Yiannis and Tim Lang. 2015 [1995]. *The Unmanageable Consumer*. 3rd ed. Los Angeles: Sage.

Galati, Frank. 1991. *The Grapes of Wrath*. London: Josef Weinberger Plays.

Galati, Frank. 2011. '*The Grapes of Wrath*: Adapting Steinbeck'. https://www.youtube.com/watch?v=UHxMXTfDOs8.

Gardner, Lynn. 26 March 2016. '*The Shepherd's Life* Review—Gentle Farming Wisdom and Animal Magic'. *The Guardian*. https://www.theguardian.com/stage/2016/mar/29/the-shepherds-life-review-theatre-by-lake-keswick-james-rebanks.

Garza, María Alicia C. 2004. 'High Crimes against the Flesh: The Embodiment of Violent Otherness in Cherríe Moraga's *Heroes and Saints*'. *Letras Femeninas* 30.1: 26–39.

Gaventa, John. 1988. 'Participatory Research in North America'. *Convergence: An International Journal of Adult Education*. 24.2–3: 19–48.

Gaventa, John and Rosemary McGee. 2010. 'Introduction: Making Change Happen—Citizen Action and National Policy Reform: Making Change Happen'. In *Citizen Action and National Policy Reform: Making Change*

Happen. Eds. John Gaventa and Rosemary McGee. London and New York: Zed Books. 1–43.

Gaventa, John and Gregory Barrett. 2012. 'Mapping the Outcomes of Citizen Engagement'. *World Development* 40.12: 2399–2410.

Geertz, Clifford. 1963. *Agricultural Involution: The Process of Ecological Change in Indonesia*. Berkeley: University of California Press.

Gill, Peter. 2001. *The York Realist*. London: Faber and Faber, Ltd.

Giagoni, Silvia. 2011. *Fields of Resistance: The Struggle of Florida's Farmworkers for Justice*. Chicago, IL: Haymarket Books.

Gioia, Paula. n.d. European Food Sovereignty Movement. http://orfc.org.uk/wp-content/uploads/2017/01/The-European-Food-Sovereignty-Movement-Paula-Gioia.pdf.

Godoy, Maria. 2017. 'Delores Huerta: The Civil Rights Icon Who Showed Farmworkers "Si Se Puede"'. [Radio Programme] NPR. https://www.npr.org/sections/thesalt/2017/09/17/551490281/dolores-huerta-the-civil-rights-icon-who-showed-farmworkers-si-se-puede.

Goldberg, Ray A. 2018. *Food Citizenship: Food Systems Advocates in an Era of Distrust*. Oxford: Oxford University Press.

Gómez-Benito, Cristóbal and Carmen Lozano. 2014. 'Constructing Food Citizenship: Theoretical Premises and Social Practices'. *Italian Sociological Review* 4.2: 135–156.

Gonzalez, Robbie. 2013. 'Think You Can Walk in a Straight Line, Blindfolded? Guess Again.' *Gizmododo*. http://io9.com/think-you-can-walk-in-a-straight-line-blindfolded-gue-1387016277.

Gouge, Melissa C. 2016. 'Human Rights in Play, Transnational Solidarity at Work: Creative Playfulness and Subversive Storytelling among the Coalition of Immokalee Workers'. *Critical Sociology* 42.6: 861–875.

Gray, Margaret. 2014. *Labor and the Locavore: The Making of a Comprehensive Food Ethic*. Berkeley and Los Angeles: University of California Press.

Greenberg, Linda Margarita. 2009. 'Learning from the Dead: Wounds, Women and Activism in Cherríe Moraga's *Heroes and Saints*'. *MELUS: Multi-Ethnic Literature of the U.S.* 34.1: 163–184.

Greenhouse, Steven. 8 November 2018. '"Some Kids Are Not Orphans Because of This": How Unions Are Keeping Workers Safe'. *The Guardian*. https://www.theguardian.com/us-news/2017/nov/08/unions-workers-safety-codes-of-conduct-florida-bangladesh.

Grossman, Marc. 2002. 'Chavez, Steinbeck: The Ties that Bind'. *The Sacramento Bee*. https://ufw.org/chavez-steinbeck-ties-bind-ufw-spokesman-longtime-cesar-chavez-press-secretary-marc-grossman-2/.

Guthman, Julie. 2014 [2004]. *Agrarian Dreams: The Paradox of Organic Farming in California*. 2nd ed. Oakland, CA: University of California Press.

Guyler-Alaniz, Marji. 2020. *FarmHer*. https://farmher.com/.

Haedicke, Susan. 2012. 'Opéra Pagaï's *Entreprise de Détournement*: Collage of Geographic, Imaginary and Discursive Spaces'. In *Theatre and the Politics of Space*. Eds. Erika Fischer-Lichte and Benjamin Wihstutz. New York: London and Routledge. 198–215.

Haedicke, Susan. 2016. 'Performing Farmscapes on Urban Streets'. *Popular Entertainment Studies* 7.1–2: 93–113.

Haedicke, Susan. 2017. '*Aroma-Home*'s Edible Stories: An Urban Community Garden Performs'. *Renewable Agriculture and Food Systems* 32.4: 1–6.

Haedicke, Susan. 2018. 'The *Aroma-Home* Community Garden Project's Democratic Narratives: Embodied Memory-Stories of Planting and Cooking'. *Public Art Dialogue* 8.1: 114–130.

Haedicke, Susan. 2020. 'Coalition of Immokalee Workers: Farmworker-led Popular Education and Performance'. *RIDE: the Journal of Applied Theatre and Performance* 25.4: 576–580.

Haedicke, Susan and Sarah Harper. 2015. '*Hope is a Wooded Time*: An Eco-Performance of Biodiversity in Discarded Geographic and Social Space'. *Performing Ethos* 5.1–2: 101–118.

Haedicke, Susan and Sarah Harper. 2018. *Who's Driving the Tractor? Conversations with Women in UK Agriculture*. https://www.youtube.com/watch?v=Jc-jwkk6p_8.

Hammelman, Colleen, et.al. 2019/2020. 'Integrated Food Systems Governance: Scaling Equitable and Transformative Food Initiatives through Scholar-Activist Engagement'. *Journal of Agriculture, Food Systems and Community Development* 9.2: 71–86.

Hammond, John L. 2014. 'Mística, Meaning and Popular Education in the Brazilian Landless Workers Movement'. *Interface: A Journal For and About Social Movements* 6.1: 372–391.

Haraway, Donna. 2008. *When Species Meet*. Minneapolis, MN: University of Minnesota Press.

Haraway, Donna. 2016. *Staying with the Trouble: Making Kin in the Chthulucene*. Durham, NC and London: Duke University Press.

Harding, James M. and Cindy Rosenthal, eds. 2006. *Restaging the Sixties: Radical Theaters and Their Legacies*. Ann Arbor: University of Michigan Press.

Harpin, Anna and Helen Nicholson, eds. 2017. *Performance and Participation: Practices, Audiences, Politics*. London: Palgrave.

Harrison, Stephan, Steve Pile and Nigel Thrift, eds. 2004. *Patterned Ground: Entanglements of Nature and Culture*. London: Reaktion Books, Ltd.

Harum, S.M. Rafael and Yelena Ogneva-Himmelberger. 2013. 'Distribution of Industrial Farms in the United States and Socioeconomic, Health, and Environmental Characteristics of Counties'. *Geography Journal*. Vol 2013. Article ID 385893. https://doi.org/10.1155/2013/385893.

Harvest of Shame. 1960. [TV programme] CBS. https://www.youtube.com/watch?v=yJTVF_dya7E.

Harvey, Fiona, Andrew Wasley, Madlen Davies and David Child. 18 July 2017. 'The Rise of Mega Farms: How the US Model of Intensive Farming Is Invading the World. *The Guardian*. https://www.theguardian.com/environment/2017/jul/18/rise-of-mega-farms-how-the-us-model-of-intensive-farming-is-invading-the-world.

Haspel, Tamer. 2017. 'Illegal immigrants Help Fuel US Farms. Does Affordable Produce Depend on Them?' *Washington Post*. https://www.washingtonpost.com/lifestyle/food/in-an-immigration-crackdown-who-will-pick-our-produce/2017/03/17/cc1c6df4-0a5d-11e7-93dc-00f9bdd74ed1_story.html?utm_term=.dbd6b9a32c14 .

Hassanein, Neva. 2003. 'Practicing Food Democracy: A Pragmatic Politics of Transformation'. *Journal of Rural Studies* 19: 77–86.

Hassanein, Neva. 2008. 'Locating Food Democracy: Theoretical and Practical Ingredients'. *Journal of Hunger and Environmental Nutrition* 3. 2–3: 286–308.

Hatanaka, Maki. 2020. 'Beyond Consuming Ethically? Food Citizens, Governance, and Sustainability'. *Journal of Rural Studies* 77: 55–62.

Hayes, Jennie. 2007. 'Fieldwork: Collaboration and Dialogue in *Focus on Farmers*'. In *Focus on Farmers: Art and Hill Farming*. Ed. Aune Head Arts. Bayley Greene, Wellington, Somerset, UK: Halsgrove. 72–80.

Heddon, Deirdre. 2017. 'The Cultivation of Entangled Listening: An Ensemble of More-Than-Human Participants'. In *Performance and Participation: Practices, Audiences, Politics*. Eds. Anna Harpin and Helen Nicholson. London: Palgrave Macmillan. 19–40.

Heidegger, Martin. 1993 [1954]. 'Building Dwelling Thinking'. In *Basic Writings*. Ed. David Farrell Krell. London: Routledge, 343–263.

Heim, Wallace. 2005. 'Navigating Voice'. In *Performing Nature: Explorations in Ecology and the Arts'*. Eds. Gabriella Giannachi and Nigel Stewart. Oxford and New York: Lang. 199–216.

Hollins, Ben and Charlotte. 2008. *The Fight for Fordhall Farm*. Much Wenlock, Shropshire: Ellingham Press.

Hooks, Bell and Amalia Mesa-Bains. 2018. *Homegrown: Engaged Cultural Criticism*. New York and London: Routledge.

Hosking, Rebecca. 2009. *A Farm for the Future*. BBC documentary. https://vimeo.com/136857929.

Hosking, Rebecca. 2021. '*A Farm for the Future*: How the Making of This Film Changed the Path of My Life'. https://www.rebeccahosking.co.uk/a-farm-for-the-future/.

Howard, Peter. 2011. *An Introduction to Landscape*. Farnham, UK: Ashgate.

Huerta, Jorge A. 1982. *Chicano Theatre: Themes and Forms*. Bilingual Press/Editorial Bilingüe: Ypsilanti, MI.
Huerta, Jorge A. 2000. *Chicano Drama: Performance, Society and Myth*. Cambridge: Cambridge University Press.
Ingold, Tim. 2007. *Lines: A Brief History*. Routledge: London and New York.
Ingold, Tim. 2011. *Being Alive: Essays on Movement, Knowledge, and Description*. Abingdon, UK: Taylor & Francis Group.
Ingold, Tim. 2011 [2000]. *The Perception of the Environment: Essays on Livelihood, Dwelling and Skill*. 2nd ed. London and New York: Routledge.
Ingold, Tim. 2015. *The Life of Lines*. Routledge: London and New York.
Intensive Farming: Advantages and Disadvantages. n.d. http://www.everythingconnects.org/intensive-farming.html .
Jackson, Julie. 2005. 'Theatrical Space and Place in the Presentational Aesthetic of Director Frank Joseph Galati'. *Theatre Topics* 15.2: 131–148.
Jackson, Shannon. 2011. *Social Works: Performing Art, Supporting Publics*. New York and London: Routledge.
Jacobs, Elizabeth. 'The Ecologies of Protest in the Theatre of Aztlán'. *Comparative American Studies: An International Journal* 10.1: 95–107.
Jhagroe, Shivant. 2019. 'Food Citizenship and Governmentality: Neo-Communitarian Food Governance in The Hague'. *Politics and Governance* 7.4: 190–201.
Johnston, Hank, ed. 2009. *Culture, Social Movements, and Protest*. Farnham, Surrey and Burlington, VT: Ashgate.
Johnson, Ian. 2017. 'Industrial Agriculture Is Driving the Sixth Mass Extinction of Life on Earth Says Leading Academic, Professor Raj Patel'. *The Independent*. https://www.independent.co.uk/environment/mass-extinction-life-on-earth-farming-industrial-agriculture-professor-raj-patel-a7914616.html.
Jones, Ffion. 2012. *Ode to Perdurance/Awdl Amser.* [YouTube] https://www.youtube.com/watch?v=6FeJK_O-HWc
Jones, Ffion. 2014. *Woollying the Boundaries: Perceptions of, and Interventions into, Upland Sheep Farming in Wales*. [PhD Thesis] https://pure.aber.ac.uk/portal/files/10625276/Jones_Ff.pdf.
Jones, Ffion. 2017. 'Agricultural Detritus and Artistic Practices: Reflecting on Animating Heritage and Reclaiming Place-Specific Narratives'. *Performance Research* 22.8: 115–126.
Jones, Gwendolyn. 2020. 'Women in Agriculture—What They Bring to the Table'. Anco Animal Nutrition Competence. https://www.anco.net/women-in-agriculture/.
Jordan, Cara. 2013. 'The Evolution of Social Sculpture in the United States: Joseph Beuys and the Work of Suzanne Lacy and Rick Lowe'. *Public Art Dialogue* 3.2. 144–167.

Jordan, John. 2019. 'The Revenge Against the Commons of the ZAD'. In *The Rural*. Ed. MyVillages. London: White Chapel Gallery and Cambridge, MA: The MIT Press. 166–171.

Joy, Leonard. 'Collective Intelligence and Quaker Practice'. The Co-Intelligence Institute. http://www.co-intelligence.org/P-QuakerCI.html and http://fieldofwheat.co.uk/artists-pages/spaces-for-listening/.

Juris, Jeffrey S. 2015. 'Embodying Protest: Culture and Performance within Social Movements'. In *Anthropology, Theatre and Development: The Transformative Potential of Performance*. Eds. Alex Flynn and Jonas Tinius. Basingstoke: Palgrave Macmillan. 227–247.

Kafer, Alison. 2013. *Feminist, Queer, Crip*. Bloomington and Indianapolis: Indiana University Press.

Kavage, Sarah. n.d. *Industrial Harvest*. http://www.industrialharvest.com/.

Kershaw, Baz and Helen Nicholson. 2011. *Research Methods in Theatre and Performance*. Edinburgh: Edinburgh University Press, Ltd.

Kester, Grant H. 2004. *Conversation Pieces: Community + Communication in Modern Art*. Los Angeles: University of California Press.

Kester, Grant H. 2011. *The One and the Many: Contemporary Collaborative Art in a Global Context*. Durham and London: Duke University Press.

Klein, Shana. 2018. 'Lady Liberty with a Tomato: A Dialogue on Art and Activism with Kat Rodriguez and The Coalition of Immokalee Farmworkers'. *Public Art Dialogue* 8.1: 98–113.

Kneafsey, Moya, et.al. 2008. *Reconnecting Consumers, Producers and Food: Exploring Alternatives*. Oxford: Berg Publishers.

Knezevic, I., et al., eds. 2017. *Nourishing Communities: From Fractured Food Systems to Transformative Pathways*. Cham, Switzerland: Springer International Publishing AG.

Krulwich, Robert. November 22, 2010. 'A Mystery: Why Can't We Walk Straight?' NPR: Morning Edition. http://www.npr.org/blogs/krulwich/2011/06/01/131050832/a-mystery-why-can-t-we-walk-straight.

Kuppers, Petra. 2007. 'Outsides: Disability Culture Nature Poetry'. *Journal of Literary Disability* 1.1: 22–33.

La Via Campesina: International Peasants Movement. n.d. https://viacampesina.org/en/.

Lacy, Suzanne, ed. 1995. *Mapping the Terrain: New Genre Public Art*. Seattle, WA: Bay Press.

Lang, Tim. 1998. 'Towards a Food Democracy'. In *Consuming Passions: Food in the Age of Anxiety*. Eds. Sian Griffiths and Jennifer Wallace. Manchester: University of Manchester Press. 13–24.

Lang, Tim. 1999. 'Food Policy for the 21st Century: Can It Be Both Radical and Reasonable?' In *For Hunger-proof Cities: Sustainable Urban Food Systems*. Eds.

Mustafa Koc, Rod MacRae, Luc J.A. Mougeot and Jennifer Welsh. Ottawa, ON, Canada: International Development Research Centre. 216–224.

Lang, Tim. 2012. 'Food Industrialisation and Food Power: Implications for Food Governance'. In *Taking Food Public: Redefining Foodways in a Changing World*. New York and London: Routledge. 11–22.

Lappé, Francis Moore. 1990. 'Food, Farming, and Democracy'. In *Our Sustainable Table*. Ed. Robert Clark. San Francisco: North Point Press. 143–159.

Latawiec, Agnieszka, et.al. 2018. 'The Environmental Costs and Benefits of High-Yield Farming'. *Nature Sustainability* 1: 477–485. https://www.nature.com/articles/s41893-018-0138-5.

'Latest Slavery Indictments Expose Exploitative Nature of H-2A "Guestworker" Program'. 5 October 2021. [Blog]. CIW. https://ciw-online.org/blog/2021/10/latest-slavery-indictments-expose-exploitive-nature-of-h-2a-guestworker-program/?emci=46f92374-db25-ec11-981f-501ac510a405&emdi=8a06d1d3-e225-ec11-981f-501ac510a405&ceid=13379203.

Legat, Allice. 2008. 'Walking Stories; Leaving Footprints'. In *Ways of Walking: Ethnography and Practice on Foot*. Eds. Tim Ingold and Jo Lee Vergunst. Farnham, Surrey, England and Burlington, VT: Ashgate. 35–49.

Lee, Wesley. 24 July 2018. 'Farmworkers Target Wendy's Executive in NYC to Demand Better Working Conditions. *Liberation*. https://www.liberationnews.org/73208-2/.

Levene, Ruth. n.d. *A Field of Wheat with Anne-Marie Culhane*. http://ruthlevene.co.uk/works/a-field-of-wheat/.

Levkoe, Charles Zalman. 2011. 'Towards Transformative Food Politics'. *Local Environment* 16.7: 687–705.

Lewis, Sue. 2012. '*Thalloo My Vea*: Narrating the Landscapes of Life in the Isle of Man'. In *Landscapes Beyond Land: Routes, Aesthetics, Narratives*. Eds. Arnar Árnason, Nicolas Ellison, Jo Vergunst and Andrew Whitehouse. Oxford and New York: Berghahn Books. 98–115.

Lewthwaite, Stephanie. 2007. 'Race, Paternalism, and "California Pastoral": Rural Rehabilitation and Mexican Labor in Greater Los Angeles'. *Agricultural History* 81.1: 1–35.

Lippard, Lucy R. 1997. *The Lure of the Local: Senses of Place in a Multicentered Society*. New York and London: The New Press.

Lippard, Lucy. 2007. 'A Frame for Farming'. In *Focus on Farmers: Art and Hill Farming*. Ed. Aune Head Arts. Bayley Greene, Wellington, Somerset, UK: Halsgrove. 38–45.

Little, Jo. 2002. *Gender and Rural Geography: Identity, Sexuality and Power in the Countryside*. London and New York: Prentice Hall.

Lorimer, Hayden. 2003. 'Telling Small Stories: Spaces of Knowledge and the Practice of Geography'. *Transactions of the Institute of British Geographers* 28.2: 197–217.

Lorimer, Hayden. 2005. 'Cultural Geography: The Busy-Ness of Being "More-Than-Representational"'. *Progress in Human Geography* 29.1: 83–94.

Lorzano-Cabedo, Carmen and Cristóbal Gómez-Benito. 2017. 'A Theoretical Model of Food Citizenship for Analysis of Social Praxis'. *Journal of Agricultural and Environmental Ethics* 30: 1–22.

Lupton, D. 1996. *Food, the Body and the Self*. London: Sage Publications Ltd.

Lymbery, Philip. 2018. *Dead Zone: Where the Wild Things Were*. London, Oxford, New York: Bloomsbury Publishing.

'Marching with Liberty'. The National Museum of American History. Smithsonian. http://americanhistory.si.edu/many-voices-one-nation/new-americans-continuing-debates-1965%E2%80%932000/marching-liberty.

Marquis, Susan L. 2017. *I Am Not a Tractor! How Florida Farmworkers Took on the Fast Food Giants and Won*. Ithaca and London: Cornell University Press.

Martin, Carol. 2013. *Dramaturgy of the Real*. Basingstoke: Palgrave Macmillan.

Martin-Shields, Charles and Wolfgang Stojetz. 2019. 'Food Security and Conflict. Empirical Challenges and Future Opportunities for Research and Policy Making on Food Security and Conflict'. *World Development* 119: 150–164.

Marx, Leo. 1991. 'The American Ideology of Space'. In *Denatured Visions: Landscape and Culture in the Twentieth Century*. Eds. Stuart Wrede and William Howard Adams. New York: Museum of Modern Art. 62–78.

Massey, Doreen. 2005. *For Space*. Los Angeles, London, New Delhi, Singapore, Washington, DC: Sage.

McCollough, Katelynn. 2014. 'Practical Farmers of Iowa Play Tackles Tricky Farmland Transfers'. *The Gazette*. https://www.hooplanow.com/articles/1140-practical-farmers-of-iowa-play-tackles-tricky-farmland-transfers.

McGovern, James and Jahana Hayes. 2021. 'Hunger Is a Political Issue. We Can Work to End It'. Civil Eats. https://civileats.com/2021/05/26/op-ed-hunger-is-a-political-decision-we-can-work-to-end-it/.

McGrath, John. 1981. *The Chevriot, the Stag and the Black, Black Oil*. London: Methuen.

McGreal, Chris. 9 March 2019. 'How America's Food Giants Swallowed the Family Farms'. *The Guardian*. https://www.theguardian.com/environment/2019/mar/09/american-food-giants-swallow-the-family-farms-iowa.

McMillen, Tracie. 2015/2016. 'Labor Gains'. *Modern Farmer*. http://ciw-online.org/wp-content/uploads/Modern-Farmer-FFP-Article.pdf.

Merleau-Ponty, Maurice. 1971 [1947]. 'Cezanne's Doubt'. In *Sense and Nonsense*. Trans. Hubert L. Dreyfus and Patricia Allen Dreyfus. Evanston, IL: Northwestern University Press. 9–25.

Merleau-Ponty, Maurice. 2012 [1945]. *Phenomenology of Perception*. Trans. Donald A. Landes. London and New York: Routledge.

Minkoff-Zern, Laura-Anne. 2014. 'Challenging the Agrarian Imaginary: Farmworker-Led Food Movements and the Potential for Farm Labor Justice'. *Human Geography* 7.1: 85–99.

Mitchell, Don. 1994. 'Landscape and Surplus Value: the Making of the Ordinary in Brentwood, CA'. *Environment and Planning D: Society and Space* 12: 7–30.

Mitchell, Don. 1998. *Cultural Geography: A Critical Introduction*. Oxford: Basil Blackwell.

Mitchell, Don. 2003. 'Dead Labor and the Political Economy of Landscape: California Living, California Dying'. In *The Handbook of Cultural Geography*. Eds. Kay Anderson, Mona Domosh, Steve Pile and Nigel Thrift. London: Sage Publications, Ltd. 233–248.

Monbiot, George. 25 March 2020. 'Covid-19 Is Nature's Wake-Up Call to Complacent Civilisation'. *The Guardian*. https://www.theguardian.com/commentisfree/2020/mar/25/covid-19-is-natures-wake-up-call-to-complacent-civilisation.

Moraga, Cherríe. 1994. *Heroes and Saints & Other Plays*. Albuquerque, NM: West End Press.

Moraga, Cherríe. 2002. *Watsonville: Some Place Not Here and Circle in the Dirt*. Albuquerque, NM: West End Press.

Mouffe, Chantal. 2000. *The Democratic Paradox*. London and New York: Verso.

MSI Integrity. 2020. 'Not Fit for Purpose: The Grand Experiment of Multi-Stakeholder Initiatives in Corporate Accountability, Human Rights and Global Governance'. https://www.msi-integrity.org/not-fit-for-purpose/.

Mulkern, Audra. 2019. *The Female Farmer Project: Documenting the Rise of Women in Agriculture*. http://www.femalefarmerproject.org/

Myers, Misha. 2011. 'Vocal Landscaping: The Theatre of Sound in Audiowalks'. In *Theatre Noise: the Sound of Performance*. Newcastle-upon-Tyne: Cambridge Scholars Publishing. 70–81.

Nancy, Jean-Luc. 2007. *Listening*. Trans. Charlotte Mandell. New York: Fordham University Press.

Nebbe, Charity and Sarah Boden. 2014. 'Women on the Farm: An Evolving Role'. Iowa Public Radio. https://www.iowapublicradio.org/post/women-farm-evolving-role-0.

Nebbe, Charity and Ben Stanton. 3 May 2018. 'AgArts Promotes Intersection of Agriculture and Art'. Interview with Mary Swander. Iowa Public Radio. https://www.iowapublicradio.org/post/agarts-promotes-intersection-agriculture-and-art#stream/0.

Nelson, Hilde Lindemann. 2001. *Damaged Identities, Narrative Repairs*. Ithaca, NY: Cornell University Press.

Nelson, Hilde Lindemann. 1995. 'Resistance and Insubordination'. *Hypatia* 10.2. 23–40.

Nelson, Robin. 2013. *Practice as Research in the Arts: Principles, Protocols, Pedagogies, Resistances*. Basingstoke: Palgrave Macmillan.
Nunn, Robert C. 2008. 'The Meeting of Actuality and Theatricality in *The Farm Show*'. In *Collective Creation, Collaboration and Devising*. Ed. Bruce Barton. Toronto: Playwrights Canada Press. 14–28.
O'Kane, Gabrielle. 2016. 'A Moveable Feast: Contemporary Relational Food Cultures Emerging from Local Food Networks'. *Appetite* 105: 218–231.
Olwig, Kenneth R. 2002. *Landscape, Nature, and the Body Politic: From Britain's Renaissance to America's New World*. Madison, WI: University of Wisconsin Press.
Olwig, Kenneth R. 2019. *The Meanings of Landscape: Essays on Place, Space, Environment and Justice*. London and New York: Routledge.
Ondaatje, Michael. 1974. 'The Clinton Special: A Film about *The Farm Show*'.
One Penny More. 2010. [YouTube] CIW video. https://www.youtube.com/watch?v=VURs-rsi_KQ&t=4s.
O'Neill, Holly, Tim Lewis and Sirin Kale. 16 August 2020. '"We're Stewards of Our Land": The Rise of Female Farmers'. *The Guardian*. https://www.theguardian.com/food/2020/aug/16/we-are-stewards-of-our-land-the-rise-of-female-farmers.
Ostrum, Elinor. 1990. *Governing the Commons: The Evolution of Collective Institutions for Collective Action*. Cambridge: Cambridge University Press.
Paolucci, Paul and Margaret Richardson. 2006. 'Sociology of Humor and a Critical Dramaturgy'. *Symbolic Interaction* 29.3: 331–348.
Pearson, 2007. *Carrlands*. http://www.carrlands.org.uk/.
Pearson, Mike. 2006. *'In Comes I': Performance, Memory and Landscape*. Exeter: University of Exeter Press.
Pearson, Mike. 2010. *Site-Specific Performance*. Basingstoke: Palgrave Macmillan.
Pearson, Mike. 2011. 'Deserted Places, Remote Voices: Performing Landscape'. In *Envisioning Landscapes, Making Worlds: Geography and the Humanities*. Eds. Stephen Daniels, Dydia DeLyser, J. Nicolas Entrikin, and Douglas Richardson. London and New York: Routledge. 280–286.
Pearson, Mike and Michael Shanks. 2001. *Theatre/Archaeology*. London: Routledge.
PerFarmance Project. https://perfarmanceproject.wordpress.com/.
PerFarmance Project West Midlands. 2016. https://blogs.warwick.ac.uk/perfarmance/.
Phillips, Patricia C. 1992. 'Temporality and Public Art'. In *Critical Issues in Public Art: Content, Context, and Controversy*. Eds. Harriet Senie and Sally Webster. Washington and London: Smithsonian Institution Press. 295–304.
Phillips, Patricia C. 2003. 'Public Art: A Renewable Resource'. In *Urban Futures: Critical Commentaries on Shaping the City*. Eds. Malcolm Miles and Tim Hall. London and New York: Routledge. 122–134.

Pitches, Jonathan. 2020. *Performing Mountains*. London: Palgrave Macmillan.

Polletta, Francesca. 2006. *It Was Like a Fever: Storytelling in Protest and Politics*. Chicago: University of Chicago Press.

Polletta, Francesca. 2008. 'Storytelling in Politics'. *Contexts* 7.4: 26–31.

Polletta, Francesca. 2009. 'Storytelling in Social Movements'. In *Culture, Social Movements, and Protest*. Ed. Hank Johnston. 1st ed. Farnham, Surrey and Burlington, VT: Ashgate. 33–53.

Polletta, Francesca. 2015. 'Characters in Politics'. *Storytelling, Self, Society* 11.1: 34–35.

Polletta, Francesca and Jessica Callahan. 2017. 'Deep Stories, Nostalgic Narratives, and Fake News: Storytelling in the Trump Era'. *American Journal of Cultural Sociology* 5.3: 392–408.

Polletta, Francesca and Pang Ching Bobby Chen. 2012. 'Narrative and Social Movements'. In *The Oxford Handbook of Cultural Sociology*. Ed. Jeffrey C. Alexander, Ronald N. Jacobs and Phillip Smith. Oxford: Oxford University Press. 487–506.

Polletta, Francesca, Pang Ching Bobby Chen, B.G. Gardner and A. Motes. 2011. 'The Sociology of Storytelling'. *Annual Review of Sociology* 37.1: 109–130.

Pretty, Jules. 2002. *Agri-Culture: Reconnecting People, Land and Nature*. London: Earthscan.

Rancière, Jacques. 2004. *The Politics of Aesthetics*. Trans. Gabriel Rockhill. London and New York: Continuum.

Rancière, Jacques. 2006. 'Problems and Transformations in Critical Art'. In *Participation*. Ed. Claire Bishop. London: Whitechapel. 83–93.

Rancière, Jacques. 2009. *The Emancipated Spectator*. Trans. Gregory Elliott. London: Verso.

Rancière, Jacques. 2010. *Dissensus: On Politics and Aesthetics*. Trans. Steven Corcoran. London and New York: Continuum.

Rape in the Fields. 2013. [TV Programme] PBS Frontline. https://www.pbs.org/wgbh/frontline/film/rape-in-the-fields/.

Ratcliffe, Thomas N. and Lori L. Hall. 2014. 'Practicing the Art of Dissent: Toward a Typology of Protest Activity in the United States'. *Humanity and Society* 38.3: 268–294.

Rebanks, James. n.d. 'One Shepherd and His Beloved Herdwick Sheep'. Interview on BBC Radio 4 Book of the Week. http://www.bbc.co.uk/programmes/articles/5mpgT5k3tSNsGBSc0kBD7fg/james-rebanks-one-shepherd-and-his-beloved-herdwick-sheep.

Rebanks, James. 16 April 2015. 'People Like Us Are Married to the Landscape'. BBC Radio 4. https://www.bbc.co.uk/sounds/play/p02p8mg7.

Rebanks, James. 2016. *The Shepherd's Life: A Tale of the Lake District*. UK and USA: Penguin Random House.

Rebanks, James. 2020. *English Pastoral: an Inheritance*. UK and US: Allen Lane/Penguin Books.

Re: Structure Lab. 2021. 'Due Diligence and Transparency'. https://static1.squarespace.com/static/6055c0601c885456ba8c962a/t/60660e41b634ac7381898670/1617301058609/ReStructureLab_DueDiligenceAndTransparencyLegislation_April2021.pdf.

Reid, Nessie. 2016. *The Milking Parlour*. https://themilkingparlour.org/.

Reid, Nessie. 2016. *The Milking Parlour Film*. https://capefarewell.com/latest/projects/the-milking-parlour.html and https://themilkingparlour.org/.

Reid, Nessie. 2016. *The Milking Parlour (Introduction Film)*. https://capefarewell.com/latest/projects/the-milking-parlour.html and https://vimeo.com/151214210.

Renz, Katie. 2005. 'Power People: An Interview with David Solnit'. *Mother Jones*. https://www.motherjones.com/politics/2005/03/people-power-interview-david-solnit/.

Reyes, Gerardo. 2018. 'Gerardo Reyes Chavez: What Is WSR?' [YouTube] CIW video. https://www.youtube.com/watch?v=yetarZVHrHg.

Rich, Frank. 1990. 'New Era for *Grapes of Wrath*'. *The New York Times*. https://www.nytimes.com/1990/03/23.theater/review-theater-new-era-for-grapes-of-wrath.html.

Robinson, Jo. 2016. *Theatre and the Rural*. Basingstoke: Palgrave Macmillan.

Roberts, Bea. 2015. *And Then Come the Nightjars*. Nick Hern Books.

Roe, Maggie and Ken Taylor, eds. 2014. *New Cultural Landscapes*. Abingdon, Oxon: Routledge.

Rojas, Laurie. 29 November 2010. 'Beuys' Concept of Social Sculpture and Relational Art Practices Today'. *Chicago Art Magazine*. http://chicagoartmagazine.com/2010/11/beuys%e2%80%99-concept-of-social-sculpture-and-relational-art-practices-today/.

Rose, Mitch. 2002. 'Landscape and Labyrinths'. *Geoforum* 33.4: 455–467.

Rose, Mitch. 2006. 'Gathering "Dreams of Presence": A Project for the Cultural Landscape'. *Environment and Planning D: Society and Space* 24: 537–554.

Roth, Clare and Charity Nebbe. 17 September 2014. '"Buying the Farm" and Buying the Farm'. Interview with Mary Swander. Iowa Public Radio. https://www.iowapublicradio.org/post/buying-farm-and-buying-farm#stream/0.

Ruscoe, Sybil. 15 May 2016. *On Your Farm: The Milking Parlour*. BBC Radio 4. https://www.bbc.co.uk/programmes/b07b9x6p.

Sachs, Carolyn E. 1983. *The Invisible Farmers: Women in Agricultural Production*. Totowa, NJ: Rowman & Allanheld.

Sachs, Carolyn E. 2018 [1996]. *Gendered Fields: Rural Women, Agriculture and Environment*. New York: Routledge.

Sachs, Carolyn E., et.al. 2016. *The Rise of Women Farmers and Sustainable Agriculture*. Iowa City: University of Iowa Press.

Sacks, Shelley. 2016. *Exchange Values*. http://exchange-values.org/.

Slavchevska, Vanya, Susan Kaaria and Sanna Liisa Taivalmaa. 2019. 'The Feminization of Agriculture: Evidence and Implications for Food and Water Security'. In *The Oxford Handbook of Food, Water and Society*. Eds. Tony Allan, et.al. Kindle Edition. Oxford: Oxford University Press: Chapter 15: n.p.

Sauer, Carl O. 1956. 'The Agency of Man on Earth'. In *Man's Role in Changing the Face of the Earth*. Ed. W. Thomas. Chicago, IL: University of Chicago Press.

Sauer, Carl O. 2007 [1925]. 'The Morphology of Landscape'. Originally published in *Geography* 2.2: 19–54. Repr. In John A. Wiens, Michael R. Moss, Monica G. Turner and David J. Mladenoff. *Foundation Papers in Landscape Ecology*. New York, NY: Columbia University Press. 36–70.

Shiva, V. 2000. *Stolen Harvest: The Hijacking of the Global Food Supply*. London: Zed Books.

Shiva, Vandana. 2015. *Who Really Feeds the World?* London: Zed Books.

Smithson, Robert. 2019. 'Untitled'. In *The Rural*. Ed. MyVillages. London: White Chapel Gallery and Cambridge, MA: The MIT Press. 208.

Sobchack, Vivian. 1999. 'Toward a Phenomenology of Nonfictional Film Experience'. In *Collecting Visible Evidence*. Eds. J.M. Gaines and M. Renov. Minneapolis, MN: University of Minnesota Press. 241–254.

Solnit, David. 2012. 'Excerpts from David Solnit's Comments During Part 1 of the Series, "Revolutionary Nonviolence?"'. [YouTube] https://www.youtube.com/watch?v=uxFj4xP4j4A.

Solnit, David. 2014. 'Arts Organizing & Strategy for Revolution'. Parts 1 and 2. [YouTube] https://www.youtube.com/watch?v=EfEKcXHaE94.

Southern Poverty Law Center. https://www.splcenter.org/.

Spirn, Anne Whiston. 1998. *The Language of Landscape*. New Haven and London. Yale University Press.

Steinbeck, John. 1992 [1939]. *Grapes of Wrath*. New York: Penguin Group.

Student Farmworker Alliance. http://www.sfalliance.org/.

Swander, Mary. 2012. *Farmscape: The Changing Environment*. North Liberty, IA: Ice Cube Press.

Swander, Mary. 2021. https://maryswander.com/.

Taber, Sarah. 16 June 2019. 'America Loves the Idea of Family Farms. That's Unfortunate'. *The Intelligencer*. https://nymag.com/intelligencer/amp/2019/06/america-loves-the-idea-of-family-farms-thats-unfortunate.html.

Taylor, Diana. 1997. *Disappearing Acts: Spectacles of Gender and Nationalism in Argentina's 'Dirty War'*. Durham and London: Duke University Press.

Taylor, Diana. 2003. *The Archive and the Repertoire: Performing Cultural Memory in the Americas*. Durham and London: Duke University Press.

The Story Behind your Food. 2018. [YouTube] CIW video. https://www.youtube.com/watch?v=Ny9YaJczqjY.

Theatre Passe Muraille. 1976. *The Farm Show*. Toronto: The Coach House Press.

Thomas, Arden Elizabeth. 2016. 'Poisoning the Mother/Land: An Ecofeminist Dramaturgy in José Rivera's *Marisol* and Cherríe Moraga's *Heroes and Saints*'. *Theatre History Studies* 35: 143–160.

Thomas, Gillian. 2018. '#MeToo Can't Fix the Workplace. Here's an Option that Can'. *The Washington Post*. https://www.washingtonpost.com/outlook/metoo-hasnt-fixed-the-workplace-heres-a-playbook-that-can/2018/11/29/bb1de86e-e762-11e8-b8dc-66cca409c180_story.html?noredirect=on&utm_term=.aed041efb823.

Thrift, Nigel. 2004. 'Performance and Performativity: A Geography of Unknown Lands'. In *A Companion to Cultural Geography*. Eds. James S. Duncan, Nuala C. Johnson, and Richard H. Schein. Malden, MA, Oxford, UK, and Carlton, Australia: Blackwell Publishing, Ltd. 121–136.

Thrift, Nigel. 2008. *Non-Representational Theory: Spaces, Politics, Affect*. London and New York: Routledge.

Thrift, Nigel and John-David Dewsbury. 2000. 'Dead Geographies—And How to Make Them Live'. *Environment and Planning D: Society and Space* 18: 411–432.

Tilley, Christopher. 2012. 'Walking the Past in the Present'. In *Landscapes Beyond Land: Routes, Aesthetics, Narratives*. Eds. Arnar Árnason, Nicolas Ellison, Jo Vergunst and Andrew Whitehouse. New Oxford: Berghahn Books. 15–32.

Tiziou, J.J. 'CIW & Fair Food'. 2017. [Photographic album] https://www.jjtiziou.net/series/ciw-fair-food.

Tuan, Yi-Fu. 1974. *Topophilia: A Study of Environmental Perception, Attitudes, and Values*. New York: Columbia University Press.

Tuck, Eve and Marcia McKenzie. 2015. *Place in Research: Theory, Methodology, and Methods*. New York and London: Routledge.

Union of Concerned Scientists. https://www.ucsusa.org/.

United Farm Workers of America. *The Official Page of the United Farmworkers of America*. https://ufw.org/.

United Farm Workers. 1986. *The Wrath of Grapes*. [YouTube] https://www.youtube.com/watch?v=Wq48o4ftL4A.

United Nations. n.d. 'Sustainable Development Goals. Goal 2: Zero Hunger'. https://www.un.org/sustainabledevelopment/hunger/.

Usmiani, Renate. 2008. 'A Success Story: Theatre Passe Muraille'. In *Collective Creation, Collaboration and Devising*. Ed. Bruce Barton. Toronto: Playwrights Canada Press. 29–37.

Valdez, Luis. 1994. *Luis Valdez—Early Works: Actos, Bernabé and Pensamiento Serpentino*. 2nd printing. Kindle edition. Houston, TX: Arte Público Press.

Van der Zee, Bibi. 4 October 2017. 'Why Factory Farming Is Not Just Cruel But a Threat to All Life on the Planet'. *The Guardian*. https://www.theguardian.com/environment/2017/oct/04/factory-farming-destructive-wasteful-cruel-says-philip-lymbery-farmageddon-author.

Vileisis, Ann. 2008. *Kitchen Literacy: How We Lost Knowledge of Where Our Food Comes From and Why We Need to Get It Back*. Washington, Covelo and London: Island Press/Shearwater Books.

Wald, Sarah. 2011. 'Visible Farmers/Invisible Workers: Locating Immigrant Labor in Food Studies'. *Food, Culture and Society* 14.1: 567–586.

Waterton, Emma. 2013. 'Landscape and Non-Representational Theories'. In *The Routledge Companion to Landscape Studies*. Eds. Peter Howard, Ian Thompson, and Emma Waterton. London and New York: Routledge. 66–75.

Wattchow, Brian. 2013. 'Landscape and a Sense of Place: a Creative Tension'. In *The Routledge Companion to Landscape Studies*. Eds. Peter Howard, Ian Thompson, and Emma Waterton. London and New York: Routledge. 66–75.

Whatmore, Sarah. 1991. *Farming Women: Gender, Work and Family Enterprise*. Basingstoke, UK: Macmillan Academic and Professional, Ltd.

Whatmore, Sarah. 1999. 'Hybrid Geographies: Rethinking the "Human" in Human Geography'. In *Human Geography Today*. Eds. Doreen Massey, John Allen, and Philip Sarre. Cambridge, UK: Polity Press. 87–96.

Whitehouse, Andrew. 2012. 'How the Land should be: Narrating Progress on Farms in Islay, Scotland'. In *Landscapes Beyond Land: Routes, Aesthetics, Narratives*. Eds. Arnar Árnason, Nicolas Ellison, Jo Vergunst, and Andrew Whitehouse. Oxford: Berghahn Books. 160–177.

Williams, Sally. 29 April 2006. 'The Fight for Fordhall Farm'. *The Telegraph*. https://www.telegraph.co.uk/foodanddrink/3325249/The-fight-for-Fordhall-Farm.html.

Wilson, Louise Ann. 2016. *Emplacing, Re-Imagining and Transforming 'Missing' Life-Events: A Feminine Sublime Approach to the Creation of Socially Engaged Scenography in Site-Specific Walking Performance in Rural Landscapes* [PhD Thesis] Lancaster University. https://eprints.lancs.ac.uk/id/eprint/85071/1/2017wilsonphdinternal.pdf.

Wilson, Louise Ann. *The Gathering/Yr Helfa*. https://louiseannwilson.com/work/the-gathering.

'With the Coalition of Immokalee Workers—Sets for a mística theatre at the March for Rights, Respect and Fair Food'. 2013. https://monacaron.com/CIW-mistica.

'With the Coalition of Immokalee Workers—Visuals for Mass Action by Florida's Farmworkers'. 2011. https://monacaron.com/artivism/coalition-immokalee-workers-visuals-mass-action-floridas-farmworkers.

'Worker-Driven Social Responsibility—Beyond the Headlines'. 2020. [Video] Worker-driven Social Responsibility Network. https://wsr-network.org/video/ .

World Bank. 2016. 'Feminization of Agriculture in the Context of Rural Transformations: What Is the Evidence?' Report no: ACS520815. http://documents.worldbank.org/curated/en/790991487093210959/pdf/ACS20815-WP-PUBLIC-Feminization-of-AgricultureWorld-BankFAO-FINAL.pdf.

World Development Movement. 2012. 'Food Sovereignty'. https://www.globaljustice.org.uk/sites/default/files/files/resources/food_sov_tricky_questions.pdf.

World Health Organization. n.d. https://www.who.int/health-topics/food-safety/.

World Health Organization. 2015. 'WHO Estimates the Global Burden of Foodborne Diseases'. https://apps.who.int/iris/bitstream/handle/10665/199350/9789241565165_eng.pdf?sequence=1.

WSR: Worker-Driven Social Responsibility Network. 2020. 'What Is WSR?' https://wsr-network.org/what-is-wsr/.

Wylie, John. 2002. 'An Essay on Ascending Glastonbury Tor'. *Geoforum* 33: 441–454.

Wylie, John. 2005. 'A Single Day's Walk: Narrating Self and Landscape on the South West Coast Path'. *Transactions of the Institute of British Geographers* 30.2: 234–247.

Wylie, John. 2007. *Landscape*. Abingdon, Oxon and New York, NY: Routledge.

Wylie, John. 2013. 'Landscape and Phenomenology'. In *The Routledge Companion to Landscape Studies*. Eds. Peter Howard, Ian Thompson, and Emma Waterton. London and New York: Routledge. 54-65.

Yeung, Bernice, Kendall Taggart and Andrew Donohue. 2014. 'California's Strawberry Industry Is Hooked on Dangerous Pesticides'. *Reveal*, from the Center for Investigative Journalism. https://www.revealnews.org/article/californias-strawberry-industry-is-hooked-on-dangerous-pesticides/

Young, Julian. 2006. 'The Fourfold'. In *The Cambridge Companion to Heidegger*. 2nd ed. Ed. Charles B. Guignon. Cambridge: Cambridge University Press. 373–392.

Zwerdling, Daniel. 2005. 'Fast-Food Deal a Big Win for Small Migrants' Group'. NPR (National Public Radio). *All Things Considered*. https://www.npr.org/templates/story/story.php?storyId=4706271.

Index

© The Editor(s) (if applicable) and The Author(s) 2021

S. C. Haedicke, *Performing Farmscapes*, Performing Landscapes,
https://doi.org/10.1007/978-3-030-82434-1

The manufacturer's authorised representative in the EU is Springer
Nature Customer Service Centre GmbH, Europaplatz 3, 69115 Heidelberg,
Germany. If you have any concerns regarding our products, please
contact ProductSafety@springernature.com

Printed and bound by CPI Group (UK) Ltd, Croydon, CR0 4YY
30/09/2026
02225456-0008